Here Be Dragons

Here Be Dragons

The Scientific Quest for Extraterrestrial Life

*

David Koerner

Simon LeVay

OXFORD
UNIVERSITY PRESS
2000

OXFORD
UNIVERSITY PRESS

Oxford New York
Athens Auckland Bangkok Bogotá Buenos Aires Calcutta
Cape Town Chennai Dar es Salaam Delhi Florence Hong Kong
Istanbul Karachi Kuala Lumpur Madrid Melbourne Mexico City
Mumbai Nairobi Paris São Paulo Singapore Taipei Tokyo
Toronto Warsaw

and associated companies in
Berlin Ibadan

Published by Oxford University Press, Inc.
198 Madison Avenue, New York, New York 10016

Library of Congress Cataloging-in-Publication Data
Koerner, David.
Here be dragons : the scientific quest for extraterrestrial life / David Koerner , Simon LeVay.
p. cm.
Includes bibliographical references and index.
ISBN 0-19-512852-4 (acid-free paper)
1. Life on other planets. 2. Life--Origin. I. LeVay, Simon. II. Title.
QB54.K54 2000
576.8'39--dc21 99–38170

1 3 5 7 9 8 6 4 2

Printed in the United States of America
on acid-free paper

Contents

Acknowledgments

We are very grateful to the following individuals who made this book possible by agreeing to interviews, and who in some cases read portions of the manuscript: Gustaf Arrhenius, Jeff Bada, Glenn Campbell, Michael Carr, Frank Carsey, Simon Conway Morris, Frank Drake, Von Eshleman, Steven Jay Gould, Bill Hoesch, Jerry Joyce, Jim Kasting, Stuart Kauffman, Phil Klass, Paul Kurtz, Ben Lane, Geoff Marcy, Chris McKay, Stanley Miller, Hans Moravec, Leslie Orgel, Tim Parker, Didier Queloz, Tom Ray, Mike Shao, Karl Stetter, Jill Tarter, Steven Weinberg, Dan Werthimer, Ned Wright, Claude Yoder, and Ben Zuckerman. We also thank Gery Allen, Paul Butler, Imre Friedmann, William Hagan, Todd Henry, Bill Langer, David Rice, and Karl Stapelfeldt for reading portions of the manuscript.

Here Be Dragons

Introduction

- At the Institute for Creation Research in San Diego, California, Duane Gish struggles to demonstrate the literal truth of the story told in Genesis: The universe was created in a week of divine labor, a week that ended with God's masterpiece, the creature made in his own image—Man.
- A few miles away, at the Salk Institute for Biological Studies, chemist Leslie Orgel has a good idea of what God was up against. After a lifetime of trying, Orgel hasn't succeeded in creating anything remotely living. But he has spawned a student who claims it can be done. Within two years.
- From the lifeless salt flats of Death Valley, planetary scientist Chris McKay digs up a spoonful of—life! McKay is impatient to try his prospecting skills on Mars.
- At the summit of Palomar Mountain, in the shadow of the mighty Hale telescope, Ben Lane fiddles with a Tinkertoy contraption of mirrors, lasers, and miniature railways. He is a junior member of a team of scientists who plan to scale up this "optical interferometer" and put it into orbit around the Sun. With it, they hope to see planets around distant stars, and maybe to find life on them.

- Paleontologists Steven Jay Gould and Simon Conway Morris tussle over the interpretation of some odd-looking, half-billion-year-old Canadian fossils. What is Life's guiding principle, they ask: Chance or Necessity?
- At the SETI (Search for Extraterrestrial Intelligence) Institute in Mountain View, California, radio astronomer Jill Tarter listens to the radio babble of the cosmos. Somewhere in the noise, she's convinced, is a message. And the champagne sits ready in her refrigerator.
- On a lonely road in central Nevada, Glenn Campbell sees a string of "golden orbs" light up the night sky. Smoke is rising from them. Do flying saucers have diesel engines?
- At Carnegie Mellon University, roboticist Hans Moravec introduces us to his latest offspring. "This is Uranus," he says proudly. "It may have to tow its brain behind on a trolley." Uranus, Moravec believes, is the ancestor of living machines that will make humans superfluous.
- At UCLA, astronomer Ned Wright measures ripples in the afterglow of the "Big Bang." Was it really just a "Little Bang"—one Creation out of many that, quite by chance, brought forth a life-friendly universe?

These people have little in common, except this: Each is responding in his or her own way—with denial, with fantasy, or with scientific derring-do—to a revolution in human thought. A revolution that knocked us off our throne at the hub of a wheeling universe and exiled us to a remote and humble planet, there to lament our downfall, or perhaps to plot a comeback.

That revolution didn't happen yesterday: it took place gradually over two millennia and more. But it had its grand moments. As when a Greek philosopher saw a curved, eclipsing shadow veil the Moon's face, and understood its meaning: Earth is round. As when Copernicus removed that round Earth from the center of all things and sent it in looping journeys around the Sun. As when Newton saw the apple fall—and saw a mechanical universe in which apples, cannonballs, and planets all moved by the same law. As when Darwin mapped our descent from four-legged, from legless, from microscopic creatures—a descent guided by chance and the struggle to survive. As when Crick and Watson reduced genetics to chemistry.

One of those moments—the discovery that the Earth orbits the

Sun—towers above the others, as far as its intellectual achievement and impact are concerned. Copernicus himself was deeply conservative. He was inclined to minimize the philosophical or religious importance of demoting the Earth to a mere planet. "Although it is not at the center of the universe," he wrote in the first volume of *De revolutionibus orbium coelestium*, published in 1543, "nevertheless its distance from the center is still insignificant, especially in relation to the sphere of fixed stars." And (without his knowledge) a preface was added to the book that made it seem as if Copernicus's theory was intended as a mere mathematical contrivance, not as an actual description of reality. But neither his own caution nor the machinations of his publisher could cushion the shock caused by the book.

That shock crossed all cultures and infiltrated every recess of human thought. "Humanity has perhaps never faced a greater challenge," wrote the poet-scientist Goethe, three centuries after the event. "For by his admission [that the Earth is not at the center of the universe], how much else did not collapse in dust and smoke: a second paradise, a world of innocence, poetry and piety, the witness of the senses, the conviction of a religious and poetic faith.... No wonder that men had no stomach for all this, that they ranged themselves in every way against such a doctrine."

How contrary to our senses, how opposite to our intuition, is the way things really are! Tycho Brahe, the brilliant Danish astronomer, expressed every human's instinctive response to Copernicus when he declared that "the body of the Earth, large, sluggish, and inapt for motion, is not to be disturbed by movement." But our senses and our intuition are the product of our species's brief existence here, at the interface of earth and air, not of a billion-year voyage across the cosmos.

Just thinking about those distances makes the mind reel. We're designed for close-in stuff—threading needles, hand-to-hand combat, throwing stones. By comparing the inputs from our two eyes, set a couple of inches apart in our heads, our brains figure in a flash what is closer and what is farther away. But no amount of staring tells us what star is closer than another.

Then Copernicus had a bright idea: If the Earth goes around the Sun once a year, he said, let's measure the positions of the stars in January, when the Earth's on one side of the Sun, and in July, when it's on the other. It would be like having eyes spaced as wide as the Earth's orbit. Surely, he thought, we'll see a difference between the two views—parallax, as we call it now. But no one could detect such a difference, even with that giant's gaze. So if the Earth truly orbits the Sun, even the nearest star must be incredibly, absurdly far away. "Consequently I shall not speak now of the vast space between the orb of Saturn and the Eighth Sphere [the fixed stars] left utterly empty of stars by this reasoning," wrote Brahe. (And why did he "not speak" of the thing he spoke of? Because there was an even more persuasive argument against Copernicus's theory: It was against the authority of Holy Writ.)

But the stars *are* incredibly, absurdly far away—even the nearest one. Proxima Centauri, an invisibly dim red star in the southern sky, has that honor: it is 40,000,000,000,000 kilometers away from us. Even if you could travel at the speed of light—which you couldn't—it would be a four-and-a-quarter-year journey. The distance to Proxima Centauri was figured out by the same method that failed the astronomers of the sixteenth century. The idea was right, but the tools weren't up to it. There were no telescopes.

And what about the farthest star? For a long time the Milky Way was the universe, and the farthest star was on the far side of it. But then, in the 1920s, came another shock, almost the equal of the one delivered by the Polish canon. Fuzzy patches in the night sky proved to be other "island universes," other galaxies. And galaxies assembled themselves into clusters, and clusters into superclusters, and these in turn receded to unfathomable distances. The farthest objects we have observed lie about 100,000,000,000,000,000,000,000 kilometers from Earth. 10^{23} kilometers, to squeeze those zeroes to a superscript. A 12-billion-year journey at the speed of light.

"The eternal silence of these infinite spaces frightens me," wrote Blaise Pascal, when only the tiniest fraction of that truth was known. What was the point of so much space? Why, if the universe was made for us, did it stretch so far beyond our reach? What could one fill it with, to take away its fearful emptiness, to give it purpose, human relevance, warmth?

Life!

The search for inhabited worlds began with Copernicus. Not that the notion hadn't been around long before. Lucretius, the Roman disciple of the Greek Atomists, spelled it out in the century before Christ: "We must therefore admit again and again," he wrote, "that elsewhere there are other gatherings of matter such as is this one which our sky holds in its eager embrace.... Now if the atoms are so abundant that all generations of living creatures could not count them, and if the same force and nature remains with the power to throw each kind of atom into its place in the same way as they have been thrown here, you must admit that in other parts of the universe there are other worlds and different races of men and species of wild beasts."[1] And the Scholastic philosophers of the Middle Ages had wrestled with the notion of "other worlds." Aristotle had denied such a possibility, for sure, but how could a Christian do so without limiting God's omnipotence?

But for Lucretius, and for all the pre-Copernican thinkers, "other worlds" were profoundly unreachable. They existed in a "beyond" that was by definition outside the limits of our senses, for everything within those limits was part of "our" world. Perhaps they were merely potential worlds—worlds that God could create (for he could do anything) but in his infinite wisdom chose not to. They were certainly not things one could see or point to. Least of all were they stars, for those were merely the lights in "our" night sky.

It was Copernicus's discovery that breathed life into the visible universe. For if the Earth revolved around the Sun, in an orbit like a planet, might not the planets in turn be like the Earth—large, solid, washed by rivers, fertile, forested, even inhabited? And hard on Copernicus's heels came Galileo with his little telescope, and saw the rocky surface of the Moon, and Jupiter's moons, and the moonlike phases of Venus, and the rings of Saturn. The planets were *places*, not points; that was the electrifying news borne by Galileo's "Starry Messenger" (*Siderius nuncius*, the title of his 1610 book). They were places one could dream of visiting or receiving visitors from.

And the stars? That was Giordano Bruno's work—to make them into "worlds." In 1584, twenty-five years before Galileo built his telescope, the mystical priest published the work whose title said it all: *De l'infinito universo e mondi—Of the infinite universe and worlds.* The stars were suns, made small and faint by distance, and there was no end to them. And around those suns orbited planets, as around our own. And on those planets was life.

Bruno died at the stake, and Galileo recanted when he was shown the rack. But there was no getting this genie back into the bottle. Kepler—he who took the perfect circles of Copernicus and bent them into impure ellipses—claimed to make out the caves where the moon people dwelled, and he wrote a whole fantastical book about their lives. And he wasn't the last astronomer to spot the work of extraterrestrials. In the eighteenth century, William Herschel—the discoverer of Uranus—saw cities, thoroughfares, and pyramids where Kepler had seen only caves. At the end of the nineteenth century came the canals of Mars. They were originally described as indistinct "channels" by an Italian, Giovanni Schiaparelli; but an American, Percival Lowell, later identified them as a complex system of artificial waterways, the work of a civilization fighting to survive on a desiccating planet.

This was what the astrononomers—the professionals—had to say. What laypeople had to say would fill many books, many genres, fact and fiction. Martians attacking the Earth or bringing otherworldly wisdom. Close encounters of the first, second, and third kinds. Contact. And governments became involved. In an astronomical folie à deux, the mad King George gave Herschel ever-larger telescopes, the better to see the cities on the Moon. Lunatics chasing Lunarians! Fast-forward two hundred years, and extraterrestrial life has become a leitmotiv of the US space program, a central justification for expenditures in the billions of dollars.

Let us make our own positions clear. As two scientists—an astronomer and a biologist—we are professional skeptics. We know of no direct evidence that a single living organism exists or has ever existed anywhere in the universe, outside of Earth. We doubt that any intelligent extraterrestrial has ever visited our planet in the past or will do so in the foreseeable future. As much as humanity may yearn for an end to its cosmic loneliness, that yearning alone will not turn gray planets green or spark chatter from silence. We must admit the possibility that we are alone forever.

But from that safe haven of skepticism, may we not venture out a way into the rough seas of speculation? For while nothing is certain, the possibilities are extraordinary—surely great enough to hazard a voyage or two. And even a negative result would be extraordinarily significant.

As a scientific discipline, the study of life in the cosmos is sometimes known as *exo*biology—"the study of life *outside*." But because our knowledge of terrestrial life is so crucial to the broader question of life in the universe, the term *cosmic biology* may be a better one. That way, terrestrial life is included, not excluded. It becomes an example, a specimen, not just an analogy or model. It gives us an "*n* of 1," as scientists like to say. Not a cornucopia, certainly, but far better than an "*n* of zero." The task, then, is to deduce from what we know of life on Earth the truly general principles of biology—principles that have shaped us in ways we now barely understand, and that apply wherever life may arise.

So terrestrial biology is one foundation of cosmic biology, and the other is astronomy, along with its infant child, space exploration. For it is astronomy's task to describe the habitats of cosmic life, and perhaps eventually to find life in those habitats. Cosmic biology means putting biology and astronomy together and forging a new science.

We will start our quest by asking: How does a Life get started? (And by the capitalized word 'Life' we mean an entire system of interdependent living things, linked by common descent, such as our own here on Earth.) What are the building blocks from which a Life is born, where do those building blocks come from, and how do they put themselves together to make the first fledgling creatures? Then, in Chapter 2, we explore the range of environments which our terrestrial Life can tolerate, to gain a feeling for the adaptibility of life and for the kinds of environments in which we may hope to find life elsewhere.

Bearing that knowledge in mind, we finally take off from Earth, in Chapter 3, to explore the solar system. We look for possible homes for life, and examine the evidence for and against the idea that a Life exists or once existed on at least one body in our solar system besides Earth. In Chapter 4, we leave the solar system behind and enter the dusty clouds where stars are born. Is it possible, we ask, that nascent stars gather the raw materials of life from the emptiness of deep space? And does the process of starbirth regularly give rise to planets too? Then, in Chapter 5, we describe the search for planets around other stars—a search that has just recently been rewarded, though in the most unexpected ways.

Complexity and evolution are the themes of Chapter 6: How do simple organisms get more complicated? Are there "rules" that guide a Life's development, here on Earth or elsewhere? And if so, do such rules tend toward complexity, interdependence, and intelligence? Or are we, as intelligent social beings, merely another example of Nature's penchant for creating oddities?

The thought that intelligence might be widespread in the universe leads inevitably to the desire to communicate—to SETI (the *s*earch for *e*xtra*t*errestrial *i*ntelligence). That enterprise is the theme of Chapter 7. Then, in the following chapter, we ask: Are aliens visiting Earth right now? What, in other words, are UFOS?

In Chapter 9, we row into a Sargasso of speculation, asking whether life as we know it is all there is. Are there Lifes based on quite different chemical principles than our own, or perhaps not based on chemistry at all? And what about 'artificial life'? Is a computer conscious? Are digital organisms alive? Will robots take over the biosphere from "squishy" creatures like ourselves?

In the final chapter, we delve briefly into the arcane world of modern physics. As we broaden the horizons of "our" world, embracing first the Moon and planets, then the stars, and finally the entire observable universe, are we seeing everything there is, or are there worlds that are truly and forever beyond our ken, as the ancients believed—worlds whose existence follows, not from direct observation, but from cosmological theories? And if so, does our world have some special status as a possible home for life?

At the root of the search for life is the tension between two ideas. One idea is what Carl Sagan called the "principle of mediocrity." This is the idea, deriving from the Copernican Revolution, that there is nothing special about our view of the universe; that what we see around us, including life, is likely to be replicated over and over, not in detail but in wonderful diversity. The universe, according to this notion, is a starlit garden to which we need only find the gate.

The opposing idea is that we do indeed have a privileged view of the universe—a view conditioned by our role as viewers. At the extreme, this idea can include Creationism, but it can also simply be the awareness that we should not infer conditions elsewhere from what we see here, because life *must* exist here. Earth is not a random sample of all planets, but a planet that had to have life, in order that we be here to think about it. If there were only one inhabited planet in the whole wide universe, we would be on it! And it would not seem unusual to us until we had searched the rest of space and failed to find our peers.

Although the Copernican viewpoint seems to favor the principle of mediocrity, there have been scientific findings that are not entirely supportive of it, at least in its most general application. In the 1930s, Edwin Hubble, perched on a rickety chair on a mountaintop behind Los Angeles, took the measure of the stars and found them to be in flight from us and from each other. Later that expansion was understood as the residue of a Big Bang that happened a mere 15 billion or so years ago. This discovery raises a difficulty that did not exist when it was possible to believe in an infinitely old universe, for if we are not special in space, we do at least seem to be special in time. Of course, one can get around this difficulty. It is possible, for example, that our cosmos is not all there is—that we are a mere momentary bubble in a froth of endless Becoming. But a timeless cosmos would have been easier.

No amount of learned discussion, only observation, can tell us which idea is closer to the truth. And that is where the excitement is today. Whether the rationale is right or wrong, the search for life in the cosmos has become Big Science and Big Engineering, involving our brightest minds and most expensive hardware.

Hic sunt dracones—"Here be dragons"—wrote the cartographers of old, to fill in the still unexplored lands of Earth. The same sense of mystery, the same lure to adventure, now colors the unexplored lands of the cosmos. Welcome to the dragon hunt.

SUGGESTED READING

Crowe, M.J. *The Extraterrestrial Life Debate, 1750–1900: The Idea of a Plurality of Worlds from Kant to Lowell.* Cambridge: Cambridge University Press, 1986.

Dick, S.J. *Plurality of Worlds: The Origins of the Extraterrestrial Life Debate from Democritus to Kant.* Cambridge: Cambridge University Press, 1982.

Dick, S.J. *The Biological Universe: The Twentieth-Century Extraterrestrial Life Debate and the Limits of Science.* Cambridge: Cambridge University Press, 1996.

Guthke, K.S. *Der Mythos der Neuzeit* (The last frontier: Imagining other worlds from the Copernican Revolution to modern science fiction). Ithaca, NY: Cornell University Press, 1990.

Goldsmith, D., ed. *The Quest for Extraterrestrial Life: A Book of Readings.* Mill Valley, CA: University Science Books, 1980.

1

Origins

How Life on Earth Began

The Museum of Creation and Earth History stands on a freeway frontage road in Santee, California, a nondescript suburb tucked among the low hills east of San Diego. We pull into the museum's forecourt and park backward in our space, thus concealing the "Darwin fish" that adorns the rear of the car. We feel a slight anxiety.

The receptionist is a pleasant elderly gentleman, who nevertheless increases our unease by asking, "You're not reporters, are you?" "No, no," we assure him truthfully, but it feels perilously close to a falsehood. A sense of transparent culpability, not experienced since Sunday school decades ago, accompanies us into the exhibition rooms. For this museum raises one of the most profound questions that humanity can ask—Where do we come from?—and offers an unambiguous answer: Scientists like ourselves have got it all wrong, and the Bible has got it exactly right.

We move through a series of small halls, named for the Six Days of Creation. In the first, light is divided from the darkness; in the second, a firmament appears; in the third, the seas are divided from the dry land. All this is visualized with the help of rather schematic artwork, to the accompaniment of classical music. Things get a little more animated on the Fourth Day with the appearance of the heavenly bodies: NASA-supplied color photographs reveal the beautiful and unexpectedly diverse faces of the planets and their moons. But the hall of the Fifth Day really comes alive: there's an aquarium with real fish and an aviary with real birds, although the birds' trills and warbles are piped in. The next hall has even more living creatures: a poisonous-looking frog, a snake (or its shed skin, at least), and even a herd of giant Madagascan cockroaches. And humans, of course—represented by their skulls. For the Sixth Day was the culmination of Creation, when God created Man in His own image and gave Man dominion over the Earth. In the next hall, the significance of the Seventh Day is explained: God rested from His labors, thus marking the end of the period in which He created the universe and the beginning of the period, still continuing today, in which He actively upholds His Creation.

The Seven Days of Creation by no means exhaust the museum's exhibits. Other rooms take us inside Noah's Ark, to the eruption of Mount St. Helens, to the Grand Canyon, to the interior of a glacier. We see the tower of Babel and pass through the Ishtar Gate, and we file down a corridor between portraits of creationists on the left, and evolutionists on the right—the saints and sinners of the Great Debate.

If the museum based its case simply on a divinely inspired faith in what the Bible says, it would be of limited interest to us. But far from it: The museum's whole purpose is to show how we can deduce the truth of the Bible story from objective study of the world around us—from science, in fact. It could properly be called the Museum of Natural Theology, for that is the name of the venerable branch of philosophy that seeks to recognize God through Reason and the study of His Works.

In making this case, of course, the museum has to face serious obstacles. Because of the detailed genealogies recounted in Genesis, the museum needs to place the beginning of all things no more than about 10,000 years in the past, while most astronomers and cosmologists claim that our universe is a million times older. The museum must compress into a mere six days processes that, in the view of the majority of scientists, took more than ten billion years. And it must make intentional what most scientists consider, in a deep sense, accidental.

The museum does not shirk this challenge. It expresses open antipathy toward Christians who try to smooth over the gulf by, for example, asserting that the "days" of Creation were metaphors for longer periods of time: that they were in fact "ages" or "eons." No, "days" were days—periods of 24 hours.

It also rejects the strategy, favored by some Christian groups, of pushing God's creative role backward in time, allowing the latter part of Creation to go forward by purely natural processes. Some believe, for example, that God lit the spark of life on Earth but allowed natural selection to do the job of getting from microbes to humans. This, in fact, was the view publicly espoused by Charles Darwin, though his private beliefs, as we shall see later, were different. With discoveries in physics and astronomy, there has been pressure to push God's role back even further. The British cosmologist Stephen Hawking, in his book *A Brief History of Time*, tells how he attended a scientific meeting at the Vatican at which the Pope admonished the conferees not to discuss what happened before the Big Bang, because that was God's province. Yet Hawking's lecture at the conference concerned the possible circularity of time, a hypothesis that, if true, would make the phrase "before the Big Bang" meaningless![1] The Museum of Creation wisely refuses to set foot on the slippery slope of biblical revisionism.

How then, does the museum propose to explain the apparent discrepancies between the Bible story and the usual teachings of science? There are several basic points. One is that, according to the museum, God created all things, including living creatures, in a fully functioning, mature state. Thus, Adam and Eve were created as normal adults, in possession of navels, for example—just as they are portrayed by Dürer and a hundred other artists. But seeing their navels, we think of umbilical cords and therefore assume that Adam and Eve were once fetuses—which they were not. And seeing the Tree of Knowledge, we assume that it was once a seed, and so on. There is the deceiving appearance of a past.

The same phenomenon, the museum argues, could explain how stars appeared in the sky on the Fourth Day, even though it would take many years for photons, traveling at the speed of light, to reach us from the newly created stars. God may have created a "functionally mature" state, including both the stars and the entire stream of photons traveling from it to us, in a single act. But seeing the photons, we naturally imagine that they originated from the star many years previously.

Of course, this line of thought can lead us into dangerous territory. Is it not equally possible that the universe is much younger than the Bible tells us? Perhaps God created the universe just a few hours or minutes ago, rather than 10,000 years ago? That vivid memory we have of reading this morning's newspaper, and every earlier memory—were they perhaps implanted in our brains to make us "functionally mature"? Do our past lives resemble those wildlife dioramas we loved as children: a couple of stuffed gazelles up front, and the rest painted on the backdrop? How to distinguish reality from illusion becomes an insoluble dilemma, once one posits the intentional creation of "mature" systems.

The museum presents a second line of argument to explain the discrepancies between creationism and conventional science. Most scientists, it argues, assume that natural processes have always occurred at the same rate. If the half-life of a radioactive isotope (the time required for half of the atoms to decay into other atoms) is now a million years, it was always a million years, because the physical laws that control radioactive decay have not changed since atoms first existed. But, the museum reminds us, we can't go back into the distant past and measure the decay rate then; therefore, the assumption of a constant rate is unjustified, and so is any finding based on that assumption, such as the age of a rock or of a fossil embedded in that rock. The Seventh Day of Creation, when God rested, was one particular time when the rates of physical processes might well have changed. Before then, light may have traveled at infinite speed, for example, thus providing an alternative explanation for how stars were seen on the day they were created.

As a matter of fact, it is not quite right to say that scientists simply assume the constancy of process rates. Many processes on Earth, such as the rate of deposition of sedimentary rocks, have been shown to vary greatly over time. Even the constancy of the great "constants," such as the strength of the gravitational force, is open to scientific debate: there are cosmologists who have developed models in which the force of gravity has changed since the Big Bang. But we can study process rates in the past with the same kinds of certainties and uncertainties with which we study them today. Some kinds of radioactive decay, for example, leave permanent tracks in rocks—rocks whose age can be estimated by other means, such as their degree of weathering or chemical transformation, their position in a sedimentary series, and so forth. One can count these tracks and thus determine whether the process of radioactive decay took place at the same rate in ancient times as it does today. In the end, our knowledge of process rates in the past is built on the mutual consistency of events that happened then, just as our knowledge of process rates today is built on the consistency of events happening now. To believe that the apparent great age of the universe is an illusion caused by decreasing process rates is really to say that time itself ran faster in the past—an assertion that belongs to metaphysics, not science.

Finally, the museum confronts the findings of conventional science by contesting the findings on science's own terms—by getting into the nitty-gritty of the data and challenging every piece of evidence, and every interpretation, that runs counter to the Bible story. Does radiometric dating of rocks at the bottom of the Grand Canyon prove them to be a billion years old? No, because if one applies the same technique to obviously recent lava flows near the canyon's rim, one gets an even earlier date—or so the museum's experts allege. Therefore the dating technique is patently untrustworthy. Did the dinosaurs go extinct 65,000,000 years ago, as the fossil record suggests? No, because dinosaurs were frequently and unambiguously sighted by humans—they called them "dragons"—as recently as the Middle Ages. Dinosaur fossils, like all other fossils, are merely the remains of the animals that drowned in Noah's Flood. Others survived, either by swimming or by being taken on board the Ark. At the museum, a painting of the Ark's interior shows what seems to be a stegosaurus lounging peaceably in its stall. The accompanying panel goes through the arithmetic to show that the Ark was plenty big enough to hold all 50,000 "kinds" of animals.

The Museum of Creation is an offshoot of the Institute for Creation Research,[2] whose offices are located in the same building, and the Institute's Senior Vice President, Duane Gish, is a Berkeley-trained biochemist who yields to no one in the discussion of scientific minutiae, whether it be the proper interpretation of an indistinct band in a sedimentary rock or the assessment of transitional forms between various fossil hominids. Woe to the "evolutionist" who agrees to debate Gish on a college campus or at a church meeting: he or she will be buried under an avalanche of particularities that collectively obliterate the conventional scientific worldview. Gish and the institute's founder, Henry Morris, have written a series of books that promote creationism as a science and label the theory of evolution a "religion"—and a false one, to boot. Of course, creationism should be taught in schools.

Where does the institute stand on extraterrestrial life? Bill Hoesch, the institute's Public Relations Officer, tells us that nonintelligent life—such as microbes—poses no problems. Creationists do not have the same need for them that "naturalists" do, since the Creator might well have chosen to put life on the Earth alone. But there is nothing to say that microbes do not exist elsewhere. With intelligent life it's a different story, especially if that life is in an "unfallen" state. In retribution for mankind's Original Sin, God put His Curse on the entire universe, Saint Paul tells us in Romans 8:22 ("For we know the whole creation groaneth and travaileth in pain together until now"). If innocent extraterrestrial creatures are laboring under this Curse, it would raise the question of whether God had acted unjustly. "That would raise some hoary theological problems for us," Hoesch says. So creationists doubt that such beings exist.

As we leave the museum and stand blinking in the afternoon sunlight, we have the sense of having torn ourselves free from a dark web of unreason, a web that might have held us in its threads until the brains were sucked out of our skulls. We feel the impulse to flag down one of those trucks hurtling by on Route 67, to breathlessly recount our trip to Eden and the saurian Ark, as if we had just returned from an alien abduction. Surely the driver would comfort us with the assurance that everything we saw and heard was an illusion?

Perhaps not. Creationism, in one form or another, is the majority worldview. Most people believe that the universe was brought into existence by divine intention, and about 40 percent of the population of the United States, according to a 1991 survey by *U.S. News and World Report*, believes in the literal truth of the Genesis story. Henry Morris, Duane Gish, and their colleagues at the Institute for Creation Research are unusual only in the fervor with which they explore the ramifications of that belief.

Of course, the Museum of Creation does represent something of an extreme position within theology. Natural theology, as practiced today, has many different perspectives on the identity of God and His role in the creation of the universe and life. For example, one school of liberal theology speaks of God as a process that is coming into being, rather than as a substance coexistent with but transcendent over matter. Process theologians, and many other liberal theologians, would not dream of contesting the date that dinosaurs went extinct, or any other scientific findings related to our origins. Our purpose in visiting the museum was not to gain an overview of current theological perspectives, but to sample the least naturalistic among them, in order to provide a contrast with what follows: The effort to explain the origins of life by natural processes.

* * * * *

Lucretius, whose belief in extraterrestrial life we mentioned in the Introduction, had an uncompromisingly naturalistic view of Creation. The gods exist, he said, but they are irrelevant. Our world assembled itself spontaneously, by the aggregation of atoms moving through a boundless extent of space. No Prime Mover was needed. Nor did the origin of life require divine intervention. "As I believe," he wrote, "no golden rope let down living things from on high into the fields ... rather, this same earth that now nourishes them from herself gave them birth."[3]

To explore this alternative vision, we visit the beachside community of La Jolla, 15 miles and a world away from the Museum of Creation. For La Jolla is home to the science-focused University of California, San Diego (UCSD) and to a host of satellite institutes and research corporations. Here we call on a group of five scientists—Jeff Bada, Stanley Miller, Gustav Arrhenius, Leslie Orgel, and Gerald Joyce—who are the closest thing to disciples of Lucretius that one may hope to find in the world today. Not that they are concerned with the entire panoply of Creation. It's that "golden rope" part that obsesses them. Can one explain the origin of life without it? It turns out to be a Herculean undertaking.

The group is called the NASA Specialized Center of Research and Training in Exobiology, one of a pair of such centers in the US. Yet the La Jolla scientists actually devote the bulk of their attention to terrestrial life—to "endobiology," if you like. "Certainly, our effort is to figure out how life began on Earth," the center's Director, Jeff Bada, tells us. "But of course that provides a model for everything else. Admittedly, we're biased by what we know about life on Earth. But I think the consensus is, if we can understand the processes that lead up to the origin of life, then given the proper conditions, it will probably be a universal process."

Jeans clad, with weather-beaten face and graying beard, Bada could be mistaken for an aging sailor. In fact, his research has taken him onto the high seas—he has sampled fluids emitted by deep-sea volcanic vents, for example. His office is no more than a few hundred feet from the ocean, at the Scripps Institution of Oceanography. And the ocean, Bada and his colleagues believe, is where life most probably originated.

"Water is best," said the first philosopher, Thales of Miletus, about six centuries before Christ. Water gave rise to all things, he claimed, including life—and certainly, water seems like the most natural place for a Life to get started: it's an excellent solvent, and there's plenty of it, on Earth at least. But water alone isn't enough. Terrestrial life is made of carbon-containing molecules—organic compounds—many of which also contain nitrogen, oxygen, and other elements. And assembling these molecules takes energy.

As mentioned earlier, Charles Darwin publicly expressed a belief that Earth's first creatures were divinely made. Perhaps he felt that he had rocked the boat sufficiently with his theory of evolution—that he would endanger the seaworthiness of his whole enterprise if he went further. But in a private letter, written in 1871, he did put forward the idea that life arose spontaneously, "in some warm little pond, with all sorts of ammonia and phosphoric salts, light, heat, electricity, etc. present."

If so, what were these chemicals and where did they come from? In 1936 the Russian chemist A.I. Oparin suggested an answer.[4] The Earth's early atmosphere, he proposed, was rich in ammonia (NH_3) and methane (CH_4), and lacked oxygen. In this "reducing" (hydrogen-donating) atmosphere, a large variety of organic molecules formed and were washed by rain into the ocean, gradually building up a "prebiotic soup." (The "soup" metaphor was actually introduced by the British geneticist J.B.S. Haldane, who had been thinking independently along the same lines.) The very first organisms, Oparin believed, were extremely simple: they didn't need to have complex metabolic pathways because everything was available in the soup—both molecules to make up their structure (such as amino acids) and molecules to break down for energy. It was the ultimate free lunch. Eventually, of course, the goodies ran out, and organisms had to learn how to make an honest living. However long that initial period may have lasted—a hundred thousand years, a million years, ten million years—it could have been no more than a moment in the Earth's history.

One of the La Jolla scientists, Stanley Miller of the UCSD Chemistry Department, tested Oparin's ideas in the laboratory. In 1952, as a graduate student working in the laboratory of Harold Urey at the University of Chicago, Miller performed an experiment that made him famous and established origin-of-life research as an experimental science.[5] He tested Oparin's hypothesis by (1) filling a flask with a "reducing atmosphere" (he chose a mixture of methane, ammonia, and hydrogen gas—H_2) over an "ocean" (a cupful or so of water) and (2) subjecting the milieu to "lightning strikes" (electrical discharges).

After a week, he analyzed what was in the water and found glycine and alanine—two of the amino acids that are building blocks of proteins. Subsequent experiments of a similar kind have revealed that a wide variety of amino acids, as well as the nucleosides that are the building blocks of DNA and RNA, are readily formed in experiments of this kind. Thus, Miller's work suggested that the building blocks of life were indeed there, free for the taking, in the Earth's primordial ocean. It was just a matter of putting them together into an organism.

Asked what it was like to have performed such a famous experiment while a graduate student, Miller tells us: "I'm sure it helped my career. But in terms of famousness—I don't know. A lot of people felt that it wasn't really science. It was attacking a problem that people didn't think about."

With the passage of the years, however, Miller has evolved from radical wunderkind to conservative defender of a possibly outmoded theory. This is on account of changing views about the composition of the Earth's early atmosphere. To understand this change, we must take a look at how scientists think the Earth and its atmosphere were created. According to current consensus, the Earth formed by the gathering together ("accretion") of smaller objects, or "planetesimals," in the disk of gas and dust orbiting the evolving Sun, 4.6 billion years ago. The main period of accretion lasted about 100 million years. During this period, the heat generated by frequent impacts kept the Earth in a molten state. For several hundred million years after that, sporadic large impacts probably prevented life from establishing itself. One such impact—by an object at least as large as Mars—is thought to have kicked a large amount of material from the Earth's mantle into orbit around the remainder of the planet. This orbiting material eventually accreted to form the Moon.

It was once generally believed that the Earth's original atmosphere was drawn directly from the disk of gas and dust from which the solar system formed. If so, it would have resembled the present atmosphere of Jupiter and Saturn, being rich in hydrogen and hydrogen-containing molecules, such as ammonia and methane, and lacking molecular oxygen (O_2). This would have been a strongly reducing atmosphere and would have been appropriate for the synthesis of organic compounds by the methods that Urey and Miller proposed. But according to the majority of contemporary researchers, the Earth was too small to attract or hold on to such a primordial atmosphere. Instead, the first atmosphere was composed of volatiles that were released from infalling planetesimals as they crashed into the magma ocean and were vaporized, or of volatiles that were outgassed from volcanoes. The main gases produced by these processes would probably have been water vapor, nitrogen, carbon monoxide (CO), carbon dioxide (CO_2), and hydrogen. Hydrogen, however, is light enough to escape from the atmosphere into space and, therefore, would not have accumulated in significant concentrations. Compounds such as methane and ammonia, if they were generated at all, would likely have been kept at very low concentrations by the destructive effect of the Sun's ultraviolet radiation.

Geochemists have had a much harder time figuring out how organic molecules could have been generated in this neutral or mildly reducing atmosphere, compared with the strongly reducing atmosphere favored by Oparin, Urey, and Miller. It's not completely impossible. Miller himself, for example, has shown that electrical discharges in a mildly reducing $CO_2/N_2/H_2O$ atmosphere can give rise to formaldehyde, and hydrogen cyanide can be produced in a similar atmosphere by ultraviolet irradiation. These compounds can go on to build larger organic molecules. Still, the process is not very efficient. "If you're going to make enough organic compounds," says Miller, "it has to be methane or ammonia, or else hydrogen and carbon dioxide and nitrogen."

So Miller tries to find ways to rescue the original scheme. He suggests to us, for example, that methane might have been released from the deep-sea volcanic vents. The vents don't release methane now, admittedly, but they might have done so, Miller says, when the Earth's atmosphere and oceans lacked oxygen. There would still be the problem of how to protect that methane from the Sun's ultraviolet radiation once it entered the atmosphere. But by happenstance, the Cornell astronomer Carl Sagan (shortly before his death in 1996), along with Chris Chyba, came up with a theory to explain how methane might have been protected.[6] They suggested that a layer of organic haze high in the Earth's atmosphere—smog, in effect—filtered out the ultraviolet radiation before it could reach the deeper layers where methane would be located. Sagan and Chyba came upon this idea because just such a smog layer does surround another body in the solar system—Saturn's largest moon, Titan (see Chapter 3).

We say "by happenstance" because Chyba and Sagan had not set out to rescue the Miller-Urey hypothesis. They wanted methane in the Earth's early atmosphere for a quite different reason. Early in the

Earth's history, the Sun was not as bright as it is at present because its nuclear fires were concentrated in a relatively small sphere near its center. Thus, the early Earth received about 30 percent less sunlight than it does now. By rights, this should have allowed the oceans to freeze solid. And once they had frozen solid, even the present-day Sun would be powerless to melt them because the ice would reflect most of the Sun's rays back into space. Chyba and Sagan were therefore looking for some way by which the early Earth might have been kept warm in spite of the Sun's faintness. A blanket of methane would do the job nicely, since methane is a greenhouse gas: it allows incoming sunlight to pass through to warm the Earth's surface, but it blocks the outgoing infrared radiation.

Ammonia, another powerful greenhouse gas, may also have played a more significant role than previously thought. According to a group led by Robert Hazen of the Carnegie Institution in Washington, D.C., who simulated the environment of the deep-sea vents in their laboratory, ammonia is generated in copious amounts from nitrogen thanks to the catalytic action of minerals present at the vents.[7] It's possible that, on the early Earth, ammonia was formed at the vents at a high enough rate that it built up to a significant concentration in the atmosphere, particularly if it was protected from the ultraviolet radiation by Sagan and Chyba's "smog." If so ammonia, like methane, could have contributed both to keeping the planet warm and to providing the raw material for the synthesis of organic compounds.

In spite of these various mechanisms by which the Milley-Urey hypothesis might be rescued, enough doubts have been sown to motivate a search for alternatives. Jeff Bada, as well as Chyba and Sagan, have explored the viability of another theory: the idea that the organic compounds in the prebiotic soup were not synthesized on Earth at all but were brought to Earth by infalling meteorites, comets, and dustgrains.

All three of these kinds of objects can contain significant amounts of organic compounds. Meteorites, especially the class of meteorite known as carbonaceous chondrites, can contain as much as 5 percent organic material. The Murchison meteorite, for example, a carbonaceous chondrite that fell in Australia in 1972, has been found to contain over seventy different amino acids, including eight of the twenty amino acids that are the building blocks of terrestrial proteins. Where these organic compounds originally came from is a topic we explore in Chapter 4.

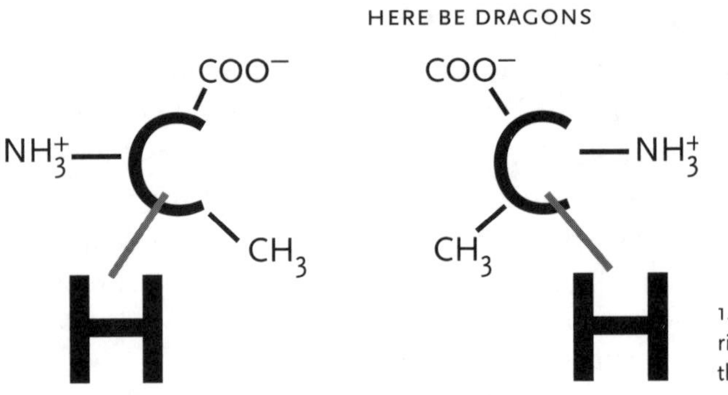

1.1 Chirality: left- and right-handed forms of the amino acid alanine.

A particularly interesting issue concerns the handedness (also called "chirality") of the organic compounds found in meteorites. Many organic compounds come in two mirror-image versions, which differ only in the three-dimensional arrangement of the bonds around one or more of the carbon atoms. The two arrangements are called, by convention, "left-handed" or "right-handed." For some reason, terrestrial life only uses amino acids of the left-handed variety. Most probably, the choice of left-handed amino acids was made because there was a small excess of left- over right-handed amino acids in the prebiotic soup. But it is very difficult to see how the local synthesis of amino acids on Earth (the hypothesis favored by Stanley Miller) could lead to more than the tiniest excess of one handedness over the other.

In the early 1970s, several groups of organic chemists reported finding an apparent excess of left-handed amino acids in carbonaceous meteorites. This opened the door to a new hypothesis: the prebiotic soup was biased toward left-handed amino acids because meteorites, comets, and dust grains imported more left handed than right-handed molecules. The findings were not terribly convincing because of the possibility that the meteorite samples had become contaminated with left-handed amino acids of terrestrial origin. In 1997, however, John Cronin and Sandra Pizzarello, of Arizona State University in Tempe, reexamined the issue by looking at the chirality of individual amino acids in the Murchison meteorite, as well as in another carbonaceous chondrite. They found that it wasn't just the familiar amino acids used in terrestrial biochemistry that were biased toward the left-handed form, so were some exotic amino acids that are never found on Earth except in meteorites. This finding seems to have disposed of the contamination issue and suggests that there was indeed a left-handed bias in the supply of amino acids that reached Earth from space. This in turn raises the question of why the extraterrestrial supply should be biased toward one chirality, a question that we will discuss in Chapter 4.

To evaluate the possible contributions of terrestrial-versus-extraterrestrial supplies of prebiotic chemicals, Bada tried to estimate how much organic material is reaching the Earth today. He suspected (as originally suggested by Edward Anders of the University of Chicago) that the bulk of the material would be brought in on very small dust grains—perhaps 50 microns in diameter or so (a micron is one-thousandth of a millimeter, or 0.0004 inches). Grains of this size don't heat up excessively as they enter the atmosphere; instead, they quickly decelerate and drift safely to the surface. To find these grains, Bada used the services of a group of French researchers, who collect and melt ice from the Antarctic—the land surface least contaminated by human activities. Starting with tons of ice, Bada ended up with micrograms (millionths of a gram—1 gram is 0.035 ounces) of material that might be of extraterrestrial origin. He then analyzed this material for the presence of organic compounds and found amino acids—not just the usual amino acids that are found in terrestrial organisms (which might represent contamination) but also amino acids that play no part in terrestrial biochemistry and that therefore are almost certainly of extraterrestrial origin.

Micrometeorites in the millimeter-size range are generally heated to incandescence as they enter the Earth's atmosphere, producing the familiar "shooting stars," and any organic freight is therefore destroyed. But there is a range of larger objects—say cabbage sized or thereabouts—that bring organic compounds to the surface intact. The surface layers of these meteorites do heat to incandescence during passage through the atmosphere, but their interiors remain cold.

Much larger objects strike the Earth's surface with such force that the heat of impact destroys any organic compounds they contain. However, large objects not uncommonly break up in the atmosphere. This happened, for example, with the "Tunguska object"—thought to have been a stony meteorite about 160 meters across—that exploded in the air over Siberia in 1908. In such cases, the resulting fragments could descend to the surface more gently.

There is another potential mechanism by which meteorites can contribute to the Earth's inventory of organic compounds, even if its own organics are destroyed on impact. Shock waves produced by the meteorite as it races through the atmosphere can provide the energy to form organic compounds, rather in the same way as did the lightning strikes in the original Miller-Urey mechanism. But like that mechanism, the shock-synthesis of organic compounds works best in a strongly reducing atmosphere and is therefore not really a means to explain how organics could form in a more plausible early atmosphere.

Bada, as well as other researchers such as Chyba and Sagan, have gone through the arithmetic to see if enough extraterrestrial organic material could have reached the early Earth to make a reasonably thick "prebiotic soup."[8] One might think that a soup of any desired thickness could be generated simply by waiting long enough: after all, there were no bacteria to eat it up. But in fact there is a process that destroys organic compounds in the ocean: this is the cycling of seawater though the deep-sea volcanic vents. Seawater in the vicinity of the midocean ridges percolates down to the magma beneath the seafloor, where it is heated and returned to the ocean through the vents. In the process, the seawater reaches a temperature of about 500°C (over 900°F)—a temperature that should be hot enough to destroy organic compounds. Bada has verified this by directly sampling the water as it passes through the vents, both in the Pacific Ocean and in the Gulf of Mexico: it is indeed "clean." Rough calculations indicate that the entire ocean circulates through the vents about once every 10 million years. Therefore one doesn't have forever to build up a prebiotic soup, and the rate at which organic compounds accumulate is crucial.

The bottom line, according to Bada, Chyba, and Sagan, is as follows. If the early Earth had a strongly reducing atmosphere, organic compounds from all sources would have built up to a steady-state concentration of about 0.1 percent—a gram of organic compounds for every kilogram of seawater. Not quite chicken broth, but in the same ballpark. If the atmosphere were neutral or mildly reducing, on the other hand, the prebiotic soup would have been a thousandfold more dilute—only about a milligram per kilogram of seawater.

Of course, no one knows how thick a soup would be needed to allow life to get rolling. And it's possible that the concentration of organic material could build up to higher levels in certain favorable locations—in drying tidal pools, for example. But still, the thousandfold difference makes the original reducing atmosphere, and the Miller-Urey mechanism, very attractive. What's more, says Bada, the Miller-Urey mechanism produces "better-quality" chemicals: a wide range of amino acids and nucleosides, whereas what comes in from space tends to be much more restricted—glycine (the simplest amino acid) and a lot of pretty useless compounds.

There is a third possibility, put forward by, of all people, a patent attorney in Munich, Germany, by the name of Günter Wächtershäuser.[9] In 1992, Wächtershäuser, who has had a lifelong passion for organic chemistry, suggested that the reduction needed to produce organic chemicals was carried out, not by gases in the atmosphere, but by inorganic chemicals at the hot deep-sea volcanic vents. Specifically, he pointed out that ferrous (iron) sulfide and hydrogen sulfide, both of which are present at the vents, constitute a powerful reducing system that might be able to convert carbon dioxide (a gas that is released from the vents) into a variety of organic compounds. These compounds, he suggested, might remain adsorbed onto the iron sulfide crystals where they were generated, thus building up high concentrations of organic molecules near the vents. If so, a prebiotic soup in the free ocean might be completely unnecessary. Wächtershäuser's "iron-sulfur world" has garnered considerable attention, particularly in the light of evidence, discussed later, that heat-loving organisms are the ancestors of all present life-forms on Earth.

According to Miller and Bada, however, Wächtershäuser's scheme doesn't hold up in practice.[10] When they (with colleagues Anthony Keefe and Gene McDonald) put together carbon dioxide, ammonia, ferrous sulfide, and hydrogen sulfide in conditions that Wächtershäuser predicted would produce amino acids, no amino acids were in fact generated. Miller and Bada believe that the proposed reactions, although theoretically feasible, require the reactants to jump over energy barriers that are not easily crossed in the absence of catalysts. "Wächtershäuser made an awful lot of noise," comments Miller, "but it doesn't work. The vents don't make organics, they destroy them."

For Gustaf Arrhenius, whose lab is also at Scripps Oceanographic, origin-of-life research is a family tradition. Gustaf is the grandson of Svante Arrhenius, a brilliant and unorthodox Swedish physicist who won a Nobel Prize in 1903 for a fundamental discovery in electrochemistry—that electricity is carried in solution by charged ions. But Svante also wrestled with profound questions about life and the universe.

"Yes, my grandfather was very interested in the origin of life," Arrhenius tells us. "Or rather, he was very uninterested. For like everyone in that period, he thought that the universe was infinite in time. Therefore, there was no reason to think that an event like the origin of life would take place just on this little speck of dust in space. It was more natural to believe that it was everywhere, that it cruised around from one place to another. Rather than worry about how life was created here on Earth, it made more sense to think how it might be transported here from elsewhere. He felt that he had hit upon a new way by which spores might be transported across space, through the pressure exerted by light. This phenomenon of light pressure, or radiation pressure, had just recently been discovered." Svante Arrhenius calculated that spores might be transported from Earth to Mars in 20 days and from the solar system to the nearest star in 9000 years.

Svante Arrhenius's theory of "panspermia" ran into a serious problem in 1910 when the French plant physiologist and radiation pioneer Paul Becquerel showed that small organisms in space would be rapidly killed by the sun's ultraviolet radiation. Later, the discovery of cosmic rays made the survival of tiny spores in space even less plausible. Yet, Svante Arrhenius held on to his theory with almost mystical fervor. He found enormously appealing the notion that all organisms in the universe were related. Perhaps life had always existed, and the question of how it might be created was therefore superfluous.

Gustaf Arrhenius, now in his seventies, betrays his Scandinavian origin with his accent, his careful choice of words, and his urbane

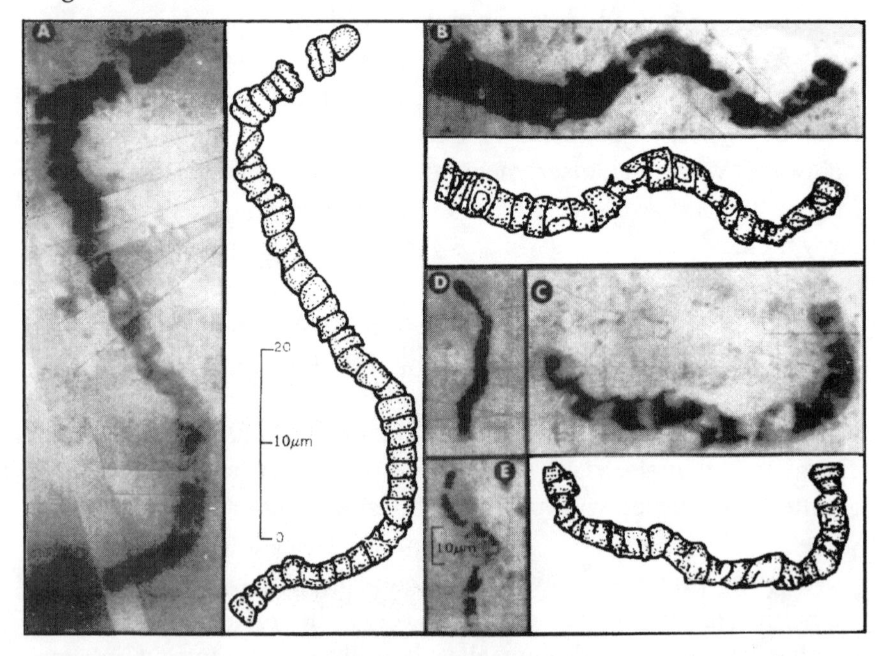

1.2 Bacterial microfossils in the 3.5-billion-year-old Apex Chert of Australia (courtesy of Bill Schopf).

good humor. And something still draws him to the chilly North, for he has recently spent time prospecting for geological specimens on the western coast of Greenland. There, he and his colleagues found what may be the most ancient remaining traces of life on earth.

Searching for the origins of life in the geological record has, of course, occupied scientists for generations. Fossils are abundant in sedimentary rocks as old as the Cambrian, a period about 500 to 600 million years ago when life diversified into all kinds of exotic and soon-to-be-discarded forms. But before the Cambrian, it gets difficult. Organisms were mostly microscopic; or, if larger, they lacked the hard body parts that readily fossilize. Even worse, the more ancient sedimentary rocks themselves become harder to find, and when they can be located, they mostly turn out to have spent time at high temperature and pressure in the depths of the Earth, an experience that plays havoc with any fossils that the rocks may have contained.

The oldest clearly recognizable fossilized organisms were found in the late 1980s in Western Australia by William Schopf of UCLA.[11] Radioactive dating of the rocks in which they were embedded gave an age of 3.5 billion years. Schopf's organisms were filamentous microbes, very much resembling certain kinds of modern cyanobacteria—bacteria that get their energy from sunlight and that release oxygen in the process. The morphological similarity between the 3.5-billion-year-old fossils and their modern counterparts suggests that saying no to evolution can pay off handsomely in the long run.

Arrhenius knew that Schopf's organisms were not likely to have been the Earth's first inhabitants—they were too complicated. So Arrhenius and research fellow Steve Mojzsis went looking in the Isua formation of West Greenland, which contains the oldest known sedimentary rocks—rocks that were laid down about 3.9 billion years ago, a mere 600 million years or so after the Earth's formation. Arrhenius and Mojzsis did not find any recognizable organisms in these rocks, which have been subjected to periods of intense heat and pressure since their deposition. But they did find microscopic grains of apatite, a form of calcium phosphate that is generally produced by living organisms rather than by geochemical processes. And within these grains were patches of carbonaceous material that might be the charred residues of ancient organisms.

To pin down the matter a little more closely, Arrhenius and Mojzsis wanted to know the isotopic composition of the carbon in these patches. If the patches were indeed derived from living organisms, they should contain relatively more ^{12}C and less ^{13}C than is found in nonliving matter. That is because the enzymes that handle carbon work better on the light ^{12}C isotope than on the heavier ^{13}C isotope. (The even heavier ^{14}C isotope does not come into the picture, because it disappears by radioactive decay over much shorter time periods than we are considering here.)

A high $^{12}C/^{13}C$ ratio was reported for Isua rocks some years ago by Manfred Schidlowski of the Max Planck Institute for Chemistry in Mainz, Germany. Schidlowski had interpreted this finding to mean that the carbon was of biogenic origin.[12] But Arrhenius and Mojzsis wanted to know specifically about the carbon in the microscopic patches. They therefore called in Mark Harrison of UCLA, an expert in a technique known as ion microprobe analysis. In a machine somewhat resembling an electron microscope, a tiny beam of positively charged cesium ions was aimed at the individual carbonate patches. The carbon atoms were vaporized and fed into a mass spectrometer, which sorted them out by mass and electric charge. It turned out that the carbon isotope ratio was indeed that expected for carbon of biogenic origin, confirming Schidlowski's interpretation.[13]

Thus, Arrhenius is confident that living organisms existed on Earth at least 3.9 billion years ago. He can say nothing about what kind of organisms they might have been, except that, given their selectivity for ^{12}C over ^{13}C, they probably had enzyme-catalyzed metabolic pathways not unlike those of organisms existing today. Earth's first organisms, therefore, should have lived long before that period.

These discoveries radically alter the scenario for the origin of life on Earth. Before the discoveries were made, it was possible to believe that the Earth lay fallow for a billion years or more, cool enough to sustain life, and rich in every nutrient required for life—yet stubbornly lifeless. What was missing, it seemed, was only the pinch of fairy dust that set everything into jangling motion. The origin of life was a probability barrier—an unlikely event that took long ages to happen, even when all circumstances seemed to favor it. But now it begins to look like the very opposite: that life arose at the very first possible moment—or even earlier!

But when was the "first possible moment"? That depends on what happened to the Earth during its first billion years, and unfortunately the history of this period is hard to reconstruct. Very large impacts, massive enough to completely vaporize any oceans that may have existed, probably continued until at least 4.3 billion years ago. Thus, any life that may have taken hold earlier should have been wiped out. Even after that period, however, major impacts apparently continued. In fact, to judge from the craters on the Moon, whose ages have been established by the study of samples returned by the Apollo astronauts, intense bombardment continued until at least 3.8 billion years ago. How can this be reconciled with an apparent origin of life well before 3.9 billion years ago?

One possible answer is that life arose on the deep-ocean floor, in an environment protected from all but the most cataclysmic impacts. This environment would also have been safe from another hazard—ice. For according to some theorists, the oceans did indeed freeze over repeatedly, to a depth of thousands of feet, only to be melted by the next all-incinerating impact. Not for nothing are the Earth's first 600 million years referred to as the Hadean Eon—the "age of Hell." Perhaps the deepest reaches of the ocean, especially the zones around the volcanic vents, did double duty as incubators and bomb shelters for the Earth's first inhabitants.

Arrhenius has a different, somewhat unconventional answer to the paradox. "I believe that the late bombardment of the Moon has nothing to do with the evolution of the Earth," he says. "It was caused by collision with objects in the Moon's own orbit—the final remains of the ring of debris that accreted to form the Moon." According to this hypothesis, none of these objects struck the Earth, and therefore the Earth may have been habitable much earlier than generally believed.

A third possibility remains unspoken—that the elder Arrhenius was right, and the Earth was indeed seeded by living organisms from other worlds. As we shall see, that could have happened otherwise than by means of light-borne spores.

Leslie Orgel, whose laboratory is at the Salk Institute, across the road from UCSD, is another distinguished elder statesman of science. Like Arrhenius, he has eschewed retirement in favor of a continued struggle with the vexatious problem of how life originated. As if worn down by the effort, the British-born chemist affects a hangdog world-weariness. When asked his opinion of some new theory, Orgel is likely to remark that "it's no worse a possibility than the others." He is emphatic only about the depths of human ignorance. Hearing that we were working on a book about life in the cosmos, Orgel comments simply: "My opinion is that we have no way of knowing anything about the probability of life in the cosmos. It could be everywhere, or we could be alone."

Yet Orgel, more than anyone else, is responsible for a profound insight about the evolution of life on Earth. He has concerned himself with the question of how one might get from a prebiotic soup of organic chemicals to an actual living system. He has concluded that the life that we're familiar with today—based on the triumvirate of DNA, RNA, and protein—was almost certainly not the original form of life on Earth. Instead, it was likely preceded by another, radically simpler form of life, in which RNA ruled alone.[14]

In today's world, the famous double helix of DNA is the almost universal repository of genetic information. RNA plays several key roles in the execution of that information, that is, in carrying out the synthesis of proteins under genetic instructions. Proteins are largely responsible for the structure of living matter and, as enzymes, for catalyzing the innumerable metabolic reactions of 'life'—including the reactions that lead to the assembly of DNA and RNA. Thus, there is a classic chicken-and-egg problem: Which came first, nucleic acids (DNA and/or RNA) or proteins? Seemingly, neither could come into being without the other, yet the idea that they arose simultaneously and by chance strains credulity.

About thirty years ago, Orgel, along with Francis Crick and Carl Woese, suggested that RNA came first. They rejected proteins as the primordial macromolecules because there was no obvious way by which proteins could replicate themselves. Nucleic acids, on the other hand, have the nucleotide base-pairing mechanism: guanine to cytosine and adenine to thymine (or to uracil in RNA). This mechanism would offer at least the theoretical possibility of replication without any assistance from proteins. Of the two nucleic acids, Orgel, Crick, and Woese favored RNA as the first-comer because of the relative ease with which the building blocks of RNA could be synthesized. In addition, several aspects of DNA biochemistry are dependent on RNA and its constituents, suggesting that RNA already existed when DNA biochemistry began.

There are three challenging requirements for an "RNA world." First, the building blocks of RNA must be available. Second, there must be a mechanism for an RNA molecule to assemble (polymerize) from those building blocks without the help of protein enzymes. And third, there must be a mechanism for RNA molecules to form copies of themselves (replicate), again without the help of proteins.

Nucleotides—the building blocks of both RNA and DNA—are composed of a base (adenine, cytosine, guanine, and thymine in DNA—

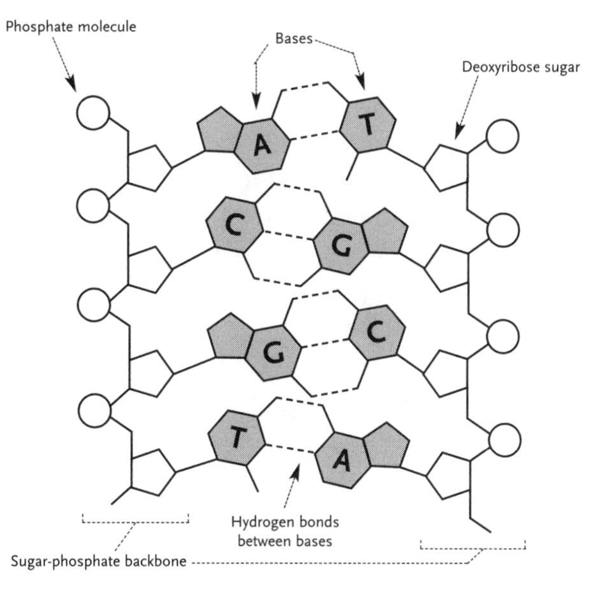

1.3 Chemical structure of DNA. In this diagram a short stretch of the familiar double helix, formed by twin sugar-phosphate backbones, has been untwisted to reveal the molecule's ladder-like structure. The four bases are adenine (A), thymine (T), cytosine (C), and guanine (G). In RNA, the sugar is ribose rather than deoxyribose, and thymine is replaced by uracil.

uracil replaces thymine in RNA) linked to a sugar (deoxyribose in DNA, ribose in RNA), which in turn is linked to a phosphate group (PO_4). (Nucleosides are nucleotides without a phosphate group.) As we've mentioned, the bases may have been available from reactions in the primordial soup, or may have been imported on micrometeorites. But making ribose and adding it on to the bases is more problematic. The difficulty is that, while it is possible to find circumstances in which ribose is made, it generally shows up as a small constituent in a mixture of many different molecules. Some of these molecules are likely to interfere with subsequent processes, just as mixing different-sized ball bearings will bring a machine grinding to a halt. Still, Orgel is reasonably optimistic that a plausible pathway will eventually be found, perhaps involving inorganic catalysts such as mineral surfaces.

Mineral surfaces might also play a role in the polymerization of RNA. James Ferris, who directs the New York Center for Studies of the Origins of Life, has managed to get ribonucleotides to assemble into short chains on the surface of a kind of clay called montmorillonite.[15] Some slightly altered ribonucleotides will form chains of an RNA-like polymer containing more than fifty bases. Again, it seems as if further research may find circumstances in which long chains of RNA will be formed.

Getting RNA to replicate is the toughest problem. For this to happen, new ribonucleotides must bind to the existing chain, following the rules of base pairing: uracil binds to adenine, cytosine to guanine.

Then, the new ribonucleotides must polymerize into a chain and separate from the original chain. The new chain, with a base sequence complementary to the original one, must now serve as the template for a second round of ribonucleotide binding, which produces a third-generation chain identical to the first. Unfortunately, although Orgel has had some limited success with the first round of copying, he has not been able to get both steps to happen. No RNA molecule has been replicated in the laboratory without the aid of protein enzymes.

"Something has to give," says Orgel. "Maybe someone will find an easy way of making nucleotides. If someone found a magic mineral which you shook up with formaldehyde or cyanide and phosphate, and out came ribose phosphate—then the whole problem would be different."

But more likely, Orgel thinks, the RNA world was not the first living system on Earth—it's just too complicated. Some simpler system had to precede it. Among the candidates for such a system is a macromolecule called peptide nucleic acid (PNA). This molecule is not known to exist in nature; it was designed by Peter Nielsen of the University of Copenhagen. PNA uses base-pairing like RNA, but has a much simpler, proteinlike backbone. Another macromolecule, called pyranosyl-RNA (pRNA), resembles RNA except that it contains a different, more abundant, version of ribose. Albert Eschenmoser, of the Swiss Federal Institute of Technology in Zurich, has coaxed pRNA into replicating under certain conditions.

But for Orgel, PNA and pRNA are still too complicated. "We want something really simple, like a polymer of aspartate and glutamate [two very similar amino acids]. Anything much more complicated than that is implausible. It's so hard to make RNA. If nothing simpler can replicate, that would be a strong argument for the existence of God."

Orgel believes that living organisms may travel from world to world, and therefore that terrestrial life may have come from another planet, as Svante Arrhenius suggested. The organisms would not travel as free-floating spores, of course. Rather, they would travel in the interior of meteorites. It's now well established that meteorites have traveled from the Moon and from Mars to Earth. Doubtless, they have traveled in the other direction too. Deep inside the meteorite, microbes would be shielded from harmful radiation, and the near-absolute-zero temperatures would keep the frozen organisms in pristine condition. "You could take *E. coli* [bacteria] and cool it rapidly to 10°K [Kelvin] and leave it for 10 billion years and then put it back in [water and] glucose, and I suspect you would have 99 percent survival," says Orgel. Takeoff and reentry are problematic, off course. But we know that the interiors of meteorites can remain cold during reentry. As for the initial impact that kicks the rock into space, Orgel draws an analogy to the circus performers who allow themselves to be shot from guns. "They use slow-burning gunpowder," he said, "so they survive the acceleration." Similarly, an impact that generated large amounts of expanding gas could accelerate a rock into space gently enough to spare the lives of the microbes within it.

Bada and Orgel are members of a NASA committee that concerns itself with the possible risks of bringing samples of extraterrestrial rocks back to Earth. Could microbes from another world set off a lethal pandemic here? The maverick British cosmologist Fred Hoyle has been saying so for years; in fact, he believes that some of the great epidemics of human history were caused by bacteria and viruses that came from space.[16]

According to Orgel, much depends on whether the Life to which those microbes belong is related to our own Life—whether one is the parent or sibling of the other, so to speak. If so, the microbes would likely be similar enough to us in their basic biochemistry that they might be able to subvert our metabolic pathways, just as pathogenic terrestrial microbes do. But if those microbes belonged to a completely independent Life, he tells us reassuringly that "they could only eat us—nothing more subtle."

Orgel wraps up our discussion on a characteristic note. "So, have we finished the origins of life? I suspect we have, haven't we? Nothing's known about it—what more is there to say? All theories are bad."

Our tour ends in a more upbeat vein, with a visit to Gerald Joyce, a molecular biologist at the Scripps Research Institute, almost next door to the Salk Institute on the bluffs overlooking the Pacific. Joyce must be about half the age of most of the other members of the NSCORT group, and he certainly plays the role of the young Turk. Where Orgel is cautious to a fault, Joyce is brazenly optimistic. A couple of years ago, he asserted that life would be created in a test tube by the end of the twentieth century.

Of course, that brings up the question: What is 'life'? Philosophers, theologians, and scientists have been torturing themselves over this for centuries. According to Joyce, there is a folk definition and a scientific definition. The folk definition of life is "that which is squishy." While open to criticism, this definition does seem to capture something about life: its plasticity, its vulnerability, and the notion that we will intuitively recognize it when we see it.

For a scientific definition, Joyce offers the product of a NASA workshop on the subject: life is a "self-sustained chemical system capable of undergoing Darwinian evolution." That means a system that can undergo reproduction, mutation, and natural selection. Of course, that definition wouldn't sit well with the people at the Museum of Creation, who don't believe that evolution happens. And later, we will have the opportunity to question other aspects of it: whether, for example, a nonchemical system such as a computer program could ever be considered "alive." But it seems like a reasonable starting point.

A "self-sustained" system, by the way, doesn't mean a closed system. To maintain itself in the face of the second law of thermodynamics—the inevitable increase in disorder or entropy in a closed system—life must use energy from outside the system. This energy could be sunlight, it could be chemical energy locked in minerals, organic compounds, or other living creatures, it could even be electricity or gravitational energy. But whatever the energy source, this requirement ensures that life can only be understood in terms of its relationship with the environment.

Joyce thinks that between life and non-life there's a sharp boundary. The origin of life wasn't the sequential appearance of gradually more 'lifelike' systems, from ones that were 1 percent living to ones that were 100 percent living. Systems gradually became more complex, certainly, but the appearance of life was at the point where such systems were able to record their phylogenetic history—which means informational macromolecules that can be replicated.

Like Orgel, Joyce works with RNA. In fact, he trained with Orgel at the Salk Institute before organizing his own lab at Scripps. But rather than focus on how RNA came to be, Joyce is engaged in the attempt to create a self-sustaining, evolving RNA system in a test tube. He wants to bring the RNA world back to life.[17]

To do this, Joyce uses a special class of RNA molecules called ribozymes. These molecules have the property not only of encoding information, but also of catalyzing enzymatic reactions—a function generally performed by proteins. To some extent, the existence of such molecules was implicit in the theory of the RNA world when it was first proposed, but it became a reality in 1983 when Thomas Cech of the

University of Colorado and Sidney Altman of Yale discovered naturally occurring RNA molecules that are able to enzymatically alter other nucleic acid molecules. These "ribozymal" RNAs are not a structurally distinct group in the sense of having, say, a chemically different backbone from other RNAs. Rather, it's simply that the particular sequence of nucleotide bases in a ribozyme causes the RNA molecule to fold up into a three-dimensional shape that is capable of catalyzing a particular reaction, just as the particular sequence of amino acids in a protein gives the molecule its particular enzymatic talent.

Joyce is after a ribozyme that can catalyze the replication of RNA—an "RNA replicase." Such a molecule, of course, could be the central player in an RNA world. If it could be found, many of the problems faced by Orgel would be circumvented. But no such ribozyme exists in the world today, as far as anyone knows. If it existed in the RNA world, as Joyce believes, it has long since gone extinct. So how to re-create it? Just trying out a bunch of randomly chosen RNA sequences doesn't seem like a workable strategy, given the numbers involved. The molecule would probably have to be at least fifty bases long to do a reasonable job, but there are about 10^{30} possible RNA sequences of this length—that's one with thirty zeroes after it. If you laid out the sequences from here to the most distant observable object in the universe, you'd have to pack a hundred billion different sequences into every millimeter of the way.

Joyce's solution is to make evolution run backward. He starts with an existing ribozyme that can, to some limited degree, carry out a reaction that is a part of the desired "replicase" function. For example, it might be able to join two short stretches of RNA. He puts this in a test tube along with a set of molecules, including DNA, protein enzymes, and nucleotides, which collectively allow the ribozyme to replicate. Within an hour, a single ribozyme molecule has multiplied a trillionfold. The trick is that Joyce arranges things so that the ribozyme's enzymatic activity is made part of the replication process. Thus, any ribozyme molecule that by chance mutates slightly, so as to perform a better enzymatic job than its peers, replicates faster. It becomes the dominant type of molecule in the tube. Thus, the mixture self-evolves in a direction set by the experimenter.

The aim is to have the ribozyme evolve to a point where it can do without the DNA and proteins. In other words, the ribozyme has to evolve backward from our present world to the RNA world. Joyce hasn't achieved that yet. But he, along with other researchers who have used similar techniques, has made significant progress. Continuous in vitro evolution, as it is called, seems to be a powerful technique for moving in a directed fashion through "sequence space"—the multidimensional realm of possible RNA sequences.[18]

So will life be created in a test tube by the end of the century, as Joyce prophesied a while back? "Sure," he says blithely, "but remember the century isn't really over till January 2001."

Figuring out how a Life gets started—whether on Earth or elsewhere—has proved to be an enormously difficult task. One wonders what Darwin would think if he could learn of all the head-spinning theories, tortuous experiments, and confusing blind alleys that have bedeviled efforts to breathe life into his "warm little pond." Would he applaud the progress, or would he throw up his hands and retreat to the notion of intentional creation? It's hard to say. But Leslie Orgel, who in many ways is the most pessimistic of the La Jolla scientists, nevertheless maintains a basic loyalty to the idea of a natural origin. "The best I can say," he says, "is that there doesn't seem to be any requirement for magic to happen."

The underlying question that remains unanswered has to do with likelihoods. The polymers that are the chemical basis of life are complicated. To assemble them by chance rolls of the dice seems to ask too much, even with an astronomical number of planets on which to roll them and 15 billion years to roll them in, for astronomical numbers are no match for hyperastronomical improbabilities. Fred Hoyle put it succinctly, when he commented that the chances of forming the simplest life by random processes were about the same as the chances that a whirlwind sweeping through a junkyard might put together a Boeing 747 airliner.[19]

The question is: Are there forces that guide the assembly process through the near-infinite maze of possibilities? Joyce's in vitro evolution hints at such a force. But what is needed, it seems, is a theory that explains the self-organization of complex systems at a more fundamental, abstract level. In Chapter 6, we will visit one place where the search for such a theory is underway.

In spite of the partial and incomplete picture that the La Jolla researchers have been able to paint, there is the thread of a story. In particular, the development of the DNA/RNA/protein world, with which we are familiar today, probably involved several successive stages of increasing complexity, with at least two major transitions: from a pre-

RNA world to the RNA world, and from the RNA world to our own world.

There is something ominous about those transitions. Did RNA "invent" DNA for example, in order to have a safer repository for its genetic information, only to have DNA "mutiny" and relegate RNA to a subservient role? These human analogies are foolish, of course, when talking about evolving chemical systems, but they're hard to avoid.

We ask Orgel whether he thinks that our own biological world will in time be taken over by the next level of complexity. We're thinking of intelligent computers or some such thing. But Orgel, ever the chemist, has his own ideas. "I claim it would happen with some weird alkaloid," he says. Alkaloids are elaborate, nitrogen-containing compounds that include, for example, heroin and caffeine. "Alkaloids look to us the way nucleotides looked to the pre-RNA world—complicated. And then one of them turns out to be a Frankenstein."

SUGGESTED READING

Chyba, C., and G.D. McDonald. "The Origin of Life in the Solar System: Current Issues." *Annual Review of Earth and Planetary Science* 23 (1995): 215–49.

De Duve, C. *Vital Dust: Life as a Cosmic Imperative*. New York: Basic Books, 1995.

Orgel, L. "The Origin of Life on the Earth." *Scientific American*, October 1994, 77–83.

2

Going to Extremes

The Habitats and Requirements for Life

It's an early April morning, and Death Valley has never looked better. El Niño's battering storms brought mayhem to coastal California, but as they charged inland, a succession of mountain ranges squeezed the wrath out of them, and they had nothing but nourishing showers for the sunken valley on California's eastern border. Desert sunflower seeds, dormant for decades, have sprouted and blossomed by the million, carpeting the valley floor with a cloth of gold.

Yet one part of the valley has resisted the urge to bloom. At Badwater, the lowest and hottest spot in the Western Hemisphere, not a living thing is to be seen: all that meets the eye is a blinding expanse of white. Water reaches this spot, to be sure. It falls as rain. It trickles down from Telescope Peak, the 11,000-foot summit that guards the valley's western wall. And the Armargosa River, usually a stony wash, but now almost fit for kayaking, brings runoff from the desert hills of western Nevada. But once at Badwater, the water turns bad. It lingers for a few days in a shallow, mineral-charged lake, and then evaporates, or is sucked into the insatiable salt.

Chris McKay strides across the lake bed, his boots crunching on the salt and gypsum crystals. With matted curly hair and beard, his shirt hanging half out of his jeans, he looks a little unkempt, and understandably so. He has driven a vanful of equipment 10 hours from his home base at NASA Ames Research Center at Moffett Field, on San Francisco Bay. He has spent one night sleeping at the side of the road and another in the campsite at Furnace Creek. His travel budget doesn't extend to motel rooms, it seems, and certainly not to a suite at the pricey Furnace Creek Inn. Besides, McKay is used to living rough. He has prospected for life in some of the world's least hospitable places: the Canadian Arctic, the wind-stripped Gobi Desert of Mongolia, the nitrate-laden expanses of Chile's Atacama Desert, and—on many occasions—the bone-dry interior of Antarctica. Most of the time, he finds what he is looking for.

Badwater is no challenge. McKay scans the featureless expanse with a practiced eye, selects a likely spot, and digs his penknife an inch into the salt. He removes what looks like a scoopful of fancy Italian ice cream: salt, for sure, but tinted in kaleidoscopic layers of pink and orange and green and black. "The pink and orange, right under the surface, that's halobacteria—microorganisms that love salt, or can deal with it at least," he says. "They have a pump in their cell membrane that kicks sodium out as fast as it comes in; it's powered by sunlight. They face death by osmosis, of course—getting all the water sucked out of them. But they fight fire with fire: they jack up the concentration of solutes inside—potassium and small organic molecules—to match the concentration of sodium chloride outside, so water doesn't move either way. They feed off whatever organic material drifts down from the sky, and they carry out a primitive kind of photosynthesis. They don't do it with chlorophyll, but with rhodopsin, the same molecule you see with, or a close relative. That's what gives them the pink color. When you fly into San Francisco and you see those pink salt-pans at the edge of the bay, that's halobacteria."

McKay points at the vivid moss-green layer beneath the halobacteria. "Those are cyanobacteria. Blue-green algae, they used to be called, but they're really bacteria. They're photosynthetic, so they want to be up at the surface, but they also need water, so they want to be deep down. They find a compromise position, about 5 or 10 millimeters beneath the surface. Although they're bacteria, they carry out photosynthesis with chlorophyll, very like green plants, splitting water and liberating oxygen. The fact that we have oxygen to breathe, that's due to the labor of these guys over 2 billion years or so. And the black stuff at the bottom, that's heterotrophic bacteria, they feed off organic compounds produced in the upper layers. There's not much oxygen down there, so they're anaerobic. Instead of breathing oxygen they breathe sulfur, so to speak. They convert sulfate to sulfide, and that's what's black."

Taking a few strides off the lake bed and across the highway, we stand under a rocky cliff, among a pile of rocks that have tumbled down from Dante's View. Here too, the first impression is of total sterility, but closer inspection reveals the presence of life. An occasional rock is encrusted with overlapping blotches of yellow or red or brown or with a branching tracery of black. "Lichen—they're fungi with algae inside," says McKay. "They can extract moisture from the air, when the humidity is seventy percent or more. The algae by themselves need a humidity of at least ninety percent, but the fungus has some kind of fancy membrane—like Gore-Tex—which lets water vapor in but doesn't let liquid water out. It's a matter of pore size."

McKay has come to Badwater to install a high-precision thermometer, which is intended to send a continuous air-temperature reading back to his office at Ames. The idea is to monitor long-term processes such as global warming. Badwater held the world's high-temperature record, at 134°F, until 1922, when a site in the Libyan desert went two degrees higher. Possibly McKay hopes to wrest the world record back for the USA.

While he and two junior colleagues attach the device to a post and program its microprocessor and transmitter, we go off in search of another ecological niche that McKay has described to us. We soon find what we're looking for on the pebbly ground north of the lake bed (see Color Plate 1). A small piece of translucent white quartzite looks as sterile as its darker neighbors, but once flipped on its back, it reveals a green underbelly—cyanobacteria again. We bring our prize back for verification. "Yup, hypolith," says McKay. "The stone acts as a moisture trap. The light comes right through, allowing photosynthesis. Quartzite stones are a major habitat for life in dry environments all over the world. They're in the Negev, the Gobi, the Atacama, the outback of Australia, even Antarctica. If it gets too dry, they go into a dehydrated vegetative state, and wait it out."

McKay is a biologist whose major interest is in identifying habitats for life on other planets, especially Mars. Until such time as his employers provide him with a ticket to the Red Planet, however, he spends much of his time investigating "extraterrestrial" habitats on

Earth—environments that in one way or another resemble conditions believed to exist on planets or moons elsewhere in the solar system. This means environments that we are likely to call extreme. But "extreme," he reminds us, is a state of mind. For halobacteria, freshwater is deadly; for the anaerobes deep in the salt, oxygen is deadly. Wherever you live, you have to pay rent. By choosing to live on dry land, for example, we humans spend an inordinate amount of energy counteracting gravity. Fish must pity us.

Death Valley challenges life with its dryness, as does Mars, where it last rained 3 billion or so years ago, according to McKay. In other respects, though, the two environments are quite different. Death Valley is hot, for example, whereas Mars is cold: rarely does any spot on Mars reach the temperature of even the coldest winter's night at Badwater.

That's what sends McKay to Antarctica, where it is both cold and dry. In some of the interior valleys, the year-round average temperature is minus 20°C, and annual precipitation is even less than in Death Valley. Yet, in these valleys, are ice-covered lakes whose lower reaches remain unfrozen year round, and in the chilly depths of these lakes is abundant life.[1]

"When I was starting this work," McKay says, "I told a colleague that the mean annual temperature was minus 20°C, and that I was going to study the lakes there. He says, 'There can't be any lakes: if the mean annual temperature is minus 20°C, they'd be frozen solid.' And he was about to explain to me how this was impossible, and I told him we'd been there, I'd been swimming in the water—theory cannot prove that there's no water there!"

It works this way: Although the mean annual temperature is far below freezing, it does rise a degree or two above the freezing point in the middle of the summer. As a result, the glaciers on the neighboring mountains melt partially, and the meltwater flows down to the ice-covered lake, where it runs under the ice. ("Ice floats," McKay reminds us. "That's probably quite an important fact in the history of life.") Once there, the water is insulated by the 5-meter icy cap. To the extent that the water freezes at all, it thickens the insulating ice layer and liberates latent heat, which further slows the freezing of the remainder. Geothermal heat, reaching the lake by conduction from the Earth's interior, also contributes to keeping the water liquid.

Although the water in these lakes is near freezing, it supports dense mats of cyanobacteria, which cover the bottom of the lake. Buoyed by the oxygen bubbles that they liberate in the course of photosynthesis, the cyanobacteria form irregular columns that rise up a few feet from the bottom, making the scene resemble a drowned city. Only about 2 percent of the Sun's light gets through the ice and reaches the cyanobacteria, but that's plenty bright enough to support photosynthesis. To McKay, in fact, the ability of some photosynthetic organisms to survive in dim light carries an important lesson for exobiology. "There are plants that photosynthesize at light levels equivalent to living at a hundred astronomical units," he says. One "astronomical unit" is the Earth's mean distance from the Sun, and a radius of a hundred astronomical units would extend more than twice as far as the Sun's most distant planet, Pluto. Therefore, McKay believes, there is no reason to think that any of our Sun's planets, or similarly placed bodies around Sun-like stars, are too dimly lit to support photosynthetic life.

Standing by one of McKay's lakes and looking around the valley in which it lies, one would take the area to be utterly devoid of life. How indeed could it be otherwise, given that there is little or no liquid water (except in the lake) and the temperature is below freezing most of the time? Yet the bare sandstone walls of the valleys are themselves richly inhabited, as was first discovered by the Hungarian-born microbiologist Imre Friedmann of Florida State University in the 1970s.[2] When Friedmann broke off a piece of sandstone and examined the broken surface, he saw three stripes: a black stripe, about 1 millimeter thick, right below the surface; a whitish stripe below that; and a green stripe about 7 to 10 millimeters deep in the rock (see Color Plate 2). The upper two stripes are made by lichens. Unlike the surface (epilithic) lichens at Death Valley, these lichens have found a livable habitat within the porous structure of the rock, so the lichens are called endoliths. The black color of the uppermost stripe is caused by pigments that shield the organisms from excessive sunlight. Without such pigment, the organisms would accumulate light-induced damage during the long periods when they are dried out and might not be able to repair that damage during the short periods when moisture is present. Additionally, the absorption of sunlight by the pigment helps to warm the rock, thus increasing the amount of time that the temperature is above freezing.

The whitish zone contains more lichens of the same species, but without pigment, since they are shielded by the layer above. Among the fungal strands of the lichens are the associated photosynthetic algae. The deepest, green zone contains algae without associated fungi: they include both eukaryotic algae and cyanobacteria. These organisms have to make do with extremely low light levels, so their growth potential is very restricted, but for the same reason, they can survive long periods of drying.

The stability of the temperature within the rock is probably an important advantage of the habitat. On the outside surface of the rock, temperatures fluctuate wildly with the vagaries of the wind: Friedmann and McKay collaborated on one set of measurements, in which they found that the surface temperature crossed the freezing point fourteen times in 42 minutes![3]. It is not surprising, then, that the rock surface is completely devoid of life. Within the rock, however, the organisms have to deal only with the day-night temperature cycle.

Although the endolithic communities seem to thrive in their chosen habitat, there are plenty of signs that they are operating in conditions near the limit of what terrestrial organisms can tolerate. In particular, the organisms are found only on the north-facing (sunlit) rock surfaces; anything else is too cold and dry. Another sign of the tough conditions is that, at many locations, all the organisms have died out. All that remain are "trace fossils"—patterns of exfoliation and leaching of the rock surface caused by the microorganisms when they were alive. These trace fossils are so distinctive that they can be recognized even millions of years after the rocks were last inhabited. This finding has implications for the search for past life on Mars (see Chapter 3).

Besides the surface lakes that McKay studies, there's at least one Antarctic lake that is completely different in nature. This is Lake Vostok, a body of water the size of Lake Ontario that lies under 3700 meters (over 12,000 feet) of ice in the Australian Antarctic Territory. It was discovered by Russian scientists, who have been drilling in this area since 1974. The liquid-water lake, which is about 500 meters deep, was detected by seismic soundings. It lies in a rift—a canyon formed by extension of the Earth's crust. It is thought that the water is kept liquid by geothermal heat and by the pressure and insulation provided by the overlying ice. The age of the lake is uncertain, but it may have been there for as much as 30 million years—that is, since before Antarctica froze over. What life, if any, it may contain is a complete mystery.

Drilling has been suspended about 100 meters above the lake, while Russian and US scientists consider how to enter the lake and take samples without contaminating it. One of the US scientists, oceanographer Frank Carsey of the Jet Propulsion Laboratory, tells us that the Russian hole is so contaminated that it may be necessary to drill an entirely new hole, but finding money to do this may be difficult. The situation at Lake Vostok, however, may be very analogous to the situation on Jupiter's moon Europa, where there is thought to be an ice-covered ocean (see Chapter 3). Thus, Lake Vostok offers a unique opportunity to develop the technology that could eventually be used to reach Europa's ocean and to search for life there.

We ask McKay whether he thinks that any life will be found in Lake Vostok. "I'd guess that there'd be no primary production," he says. "You may read a headline saying 'Life in Antarctic Lake,' but it'll just be scavengers that eat spores or whatever else is trickling in. To me a Life is an environment in which someone makes a living, and someone else eats them. Someone's got to be making a living, or it's not a Life, it's just a landfill."

By "primary production" McKay means the initial synthesis of organic compounds such as amino acids, sugars, and so on, by organisms at the bottom of the food chain. Nearly all primary production, McKay says, relies on photosynthesis and therefore usually occurs at the surface of the land or ocean, or no deeper than light can penetrate. Some primary producers labor in darkness, to be sure. At the deep-sea vents, for example, there are bacteria that get their energy by converting hydrogen sulfide, a gas that is emitted by the vents, into sulfate. But to do this, they need oxygen, and all free oxygen on Earth comes eventually from the activities of oxygen-liberating photosynthetic organisms such as cyanobacteria and green plants. "Pity the life-forms that survive on a world in which primary productivity is only chemosynthesis," says McKay. "Life needs sunlight."

While Chris McKay focuses primarily on cold, dry environments, Karl Stetter seeks hot, wet ones. Stetter, a German microbiologist on the faculty of the University of Regensburg, is another tireless globe-trotter. Given their different interests, the two men's paths cross only rarely, but McKay tells us that they did meet a few years back at McMurdo Station, the entryway to the Antarctic. McKay was heading for the dry interior valleys, while Stetter was hunting for organisms on Mount Erebus, the 4000-meter (13,000-ft) volcano that rises from the Ross Sea not far from McMurdo.

We catch up with Stetter at a less exotic location, a Los Angeles hotel. He is a frequent visitor to UCLA, the home base of fossil-microbe hunter Bill Schopf (see Chapter 1). Recalling his meeting with McKay, Stetter adds a detail that McKay didn't bother to mention—that McKay had a broken leg at the time. Of himself, Stetter says: "I'm not the wild adventurer that people think. I'm a very careful person. But I have the privilege to visit parts of the world that other people would never see—like the summit of Mt. Erebus, or the mid-Atlantic ridge, four kilometers beneath the ocean surface—so the adventures happen by themselves."

Stetter looks to be in his fifties and has thinning, sandy hair. While speaking, he waves his hands wildly in the air or stabs a finger at his questioner; his facial features go through an unending sequence of contortions, as if struggling with the fine points of the English language.

Stetter seeks *hyperthermophiles,* the "lovers of beyond hot."[4] To such organisms the ground surface at Badwater, which can reach a foot-blistering 60°C (140°F) on a summer afternoon, would be paralyzingly cold. Hyperthermophiles thrive at 80°, 90°, and 100°C. In the ocean depths, where the pressure of overlying seawater keeps the water from boiling, hyperthermophiles may thrive at even higher temperatures. The current record holder, a bug called *Pyrolobus fumarii,* was found by Stetter and the late Holger Jannasch, of the Woods Hole Oceanographic Institution, at a hot vent on the mid-Atlantic ridge, under 3650 meters of seawater. It grows only at temperatures between 90° and 113°C and can survive at 120°C (248°F) for at least an hour. Cool it below 90°C, a temperature that in humans causes third-degree burns, and it stops growing for lack of heat.

Stetter and others have found hyperthermophiles wherever the temperatures are appropriate: in the boiling hot springs of Yellowstone caldera and other land volcanoes, deep under Arctic permafrost, in shallow submarine hot springs, at deep-ocean volcanic vents, even in the hot effluent from nuclear power stations and smoldering coal-refuse piles. Since hot water, and especially hot seawater, can hold very little dissolved oxygen, it is no surprise that many hyperthermophiles are anaerobic: they do without oxygen and are often poisoned by it. At lower temperatures, however, when the organisms go into a state of suspended animation, they can tolerate oxygen. It is in this state, presumably, that they are carried from one isolated hot-water habitat to another.

It was the isolation of Mount Erebus that intrigued Stetter. "Mount Erebus is a very active volcano," he says, "but could there be hyperthermophiles in a continent that's been frozen for 30 million years? Even where the fumaroles (steam and gas vents) come up, they build up huge ice caves. It's quite spectacular. There's no liquid water on the surface, but there's a layer of water in the soil, heated from below. There we found hyperthermophiles."

Besides his skills as an explorer, Stetter says he has been credited with "green thumbs" for his ability to make exotic organisms grow in laboratory culture. "One word I don't like is *non-culturable*, he says. "It's a challenge. We can't use the techniques of the great old men like Louis Pasteur and Robert Koch. For example, we can't isolate organisms by plating them onto culture medium [like agar]. The medium deteriorates at those high temperatures. Instead, I pick out single organisms under the microscope, using 'optical tweezers,' a highly focused infrared laser beam."

"I didn't want to have a curiosity shop. I wanted to do it on a bigger scale. I call it my 'witches' kitchen,'" Stetter says, alluding to a scene from Goethe's *Faust*. "So I developed fermenters, up to 300 liters capacity. The challenge is to bring hydrogen to the organisms at temperatures above 100°C—the end product may be hydrogen sulfide, which corrodes even stainless steel within days. One fermenter was looking almost like modern art. So now we use titanium."

One central question that Stetter and many other researchers have attempted to answer is: What is it about hyperthermophiles that allows them to survive at temperatures that are lethal to most other organisms? And is there any upper limit to the temperatures at which life might exist? Just like McKay's research, therefore, Stetter's work on hyperthermophiles helps define the cosmic habitats in which life might be found.

The key to survival at a high temperature, it turns out, is not a single clever trick but a large number of minor differences that collectively allow the molecular constituents of hyperthermophiles to stay intact as the heat stress increases. In general, biological macromolecules such as proteins, nucleic acids, and lipids (fats and oils) are relatively rigid at low temperatures and become progressively more flexible or floppy as the temperature increases. There is some point along this rigid-flexible continuum where a particular kind of molecule does its job best: at too low a temperature it is incapable of making the subtle adjustments in molecular shape that allow it to interact with other molecules, while at too high a temperature it becomes so unstable that it spends little time in anything close to the proper shape and may ultimately break down irreversibly. In hyperthermophiles, the correct degree of flexibility is achieved at a much higher temperature than is the case with conventional organisms.

How is the flexibility of macromolecules adjusted? In the case of proteins, such as enzymes, the basic answer was found by the pioneering British molecular biologist Max Perutz in the 1970s. Proteins consist of one or more polypeptides (chains of amino acids) folded into knots that look random but are in fact amazingly precise. What stops the knot from coming unraveled is the presence of cross-bridges between pairs of amino acids in places where two stretches of the chain happen to come close to each other. These bridges are of various kinds, but the kind that seems to be important for heat stability is the ion pair, in which an amino acid carrying a negatively charged side group (e.g., aspartic acid) is attracted to an amino acid with a positively charged side group (e.g., arginine).

Two groups of researchers—one at the University of Regensburg led by Stetter's colleague Rainer Jaenicke, and the other at the University of Sheffield, England, led by David Rice—have used the laborious technique of x-ray crystallography to determine the three-dimensional shape of enzymes taken from hyperthermophiles and have compared the results with those from conventional organisms.[5] Both groups find that the enzymes from heat-loving organisms are much richer in ion pairs—they possess almost twice as many as their conventional counterparts. Furthermore, the ion pairs are linked into networks girdling the molecule; these networks are thought to hold the protein in a relatively compact shape, helping it resist heat-induced motions (see Color Plate 3). So far, however, attemps to manipulate the heat-stability of enzymes by varying the number and location of ion pairs have been disappointing.

Somewhat comparable differences are seen between the lipids of hyperthermophile and conventional organisms. But the most intriguing molecular differences concern DNA. Organisms can adjust the stability of their DNA by varying the proportions of the two kinds of base pairs, adenine-thymine (A-T) and guanine-cytosine (G-C). The G-C pair sticks together by means of three "hydrogen bonds," the A-T pair by only two. Hyperthermophilic organisms tend to have a high proportion of G-C pairs, compared with conventional organisms, and thus the two strands of the double helix are less easily pulled apart by random thermal motions.

In addition, however, all hyperthermophile organisms (and no organisms that live at conventional temperatures) possess an enzyme called "reverse gyrase," discovered by Akihiko Kikuchi and Keiko Asai of the Mitsubishi-Kasei Institute of Life Sciences in Tokyo,

Japan.[6] To understand what this enzyme does, one has to be aware that in prokaryotes (single-cell organisms without nuclei) most of the cell's DNA takes the form of a very long double-helical molecule that is joined at the ends to form a closed loop. Because the loop is closed, the total number of turns of the double helix, as one goes once around the loop, is topologically fixed: unwinding the helix at one location must "overwind" it at other locations, and vice versa. The reverse-gyrase enzyme, however, is able to cut the loop, add extra turns to the helix, and reseal it, leaving the DNA in a permanently overwound condition. The overwinding tightens the double helix, thus resisting the tendency of the molecule to become too flexible at high temperatures.

Microorganisms that live at conventional temperatures, such as the much-studied *E. coli*, don't possess reverse gyrase, but they do possess an enzyme called "gyrase," that has exactly the opposite function: it "underwinds" DNA, leaving it in a more floppy condition. (Gyrase was discovered first: that's the only reason that reverse gyrase is called "reverse.") *E. coli* switches on its gyrase enzyme when it is exposed to cold. Thus, we can begin to see how microorganisms can keep their DNA in an optimal state of flexibility in the face of large differences in temperature.

So what is the upper temperature limit for terrestrial life? According to Stetter, life could exist at temperatures well above 120°C. The molecular backbones of proteins and DNA remain intact up to temperatures of about 180°C. Some of the small molecules that play a vital role in metabolism, such as the universal energy supplier adenosine triphosphate (ATP), break down at lower temperatures, so organisms living at 180°C—if any do—would have to find a more heat-stable substitute. When pressed, Stetter suggests 150°C (302°F) as the probable upper limit for life "as we know it."

Many hyperthermophiles lead a biochemical lifestyle that seems very alien to our own, yet when we study these lifestyles in detail, we can get a sense of the underlying similarity of all living processes. Take a deep-sea vent organism called *Methanopyrus*—actually a genus, or collection of closely related species. It lives in the water-permeable rocky walls of the chimneys through which the hot-vent fluids emerge into the ocean. Stetter has found that as many as a hundred million *Methanopyrus* organisms can inhabit a single gram of the rock. (Still, that's not as densely populated as ordinary garden soil, which contains more than a billion microorganisms per gram.)

Methanopyrus makes a living by taking molecular hydrogen (H_2) and carbon dioxide (CO_2), both of which are present in the vent fluids, and making methane (CH_4) and water. The reaction can be written like this:

$$4H_2 + CO_2 \rightarrow CH_4 + 2H_2O$$

This is an oxidation-reduction, or electron-transfer, reaction: electrons are being stripped from the hydrogen gas and donated to the carbon dioxide, thus reducing it. Protons come along for the ride, so the carbon atom ends up surrounded by uncharged hydrogen atoms rather than by negatively charged electrons. But the protons are pretty irrelevant from an energy-budget point of view because in any watery environment they are free for the taking—they are formed spontaneously by the dissociation of water molecules. Electrons, on the other hand, do not exist as free particles in water; rather, they must be transferred directly from donor to acceptor molecules. That is the reason that oxidation-reduction reactions in aqueous solution are usually defined as electron transfers rather than as hydrogen transfers.

In this sense, beasts as different as *Methanopyrus* and *Homo sapiens* make their living in the same way. Humans survive by burning simple foodstuffs such as glucose ($C_6H_{12}O_6$) by the following reaction:

$$C_6H_{12}O_6 + 6O_2 \rightarrow 6CO_2 + 6H_2O$$

in which electrons are transferred from the glucose to the oxygen. Organisms can in principle derive energy from any electron-transfer reaction, but only by running the reaction in the downhill direction—that is, the direction in which the free energy of the entire system (the energy available for doing work) is reduced. This, of course, is mandated by the second law of thermodynamics. The energy difference between the two states may be released as heat or, as we'll see in a moment, may be put to more constructive uses.

Everett Shock of Washington University, St. Louis, has studied the chemical system in which *Methanopyrus* lives.[7] The fluid rising through the vent is at a temperature of about 400°c. At this high temperature, the carbon dioxide and hydrogen are in equilibrium: no energy is to be gained by a reaction between them. As the fluid mixes with cold seawater, however, the temperature drops to a range (60–250°c) in which the gases are out of equilibrium: the conversion to methane lowers the free energy of the system. As it cools even further, below 60°c, the mixture of hydrogen and carbon dioxide is again in the lower-energy state, and the conversion to methane is no longer favored. Only if *Methanopyrus* locates itself in the middle-temperature region can it extract energy from the conversion. Thus, the heat-tolerant nature of *Methanopyrus* is absolutely crucial to its ability to utilize the $CO_2 \longrightarrow CH_4$ reaction as an energy source.

Stetter mentions another crucial variable—pressure. "The reaction used by *Methanopyrus* has a negative reaction volume, meaning the volume of the dissolved gases is less at the end of the reaction. Five volumes of gas are reduced to one volume of gas. Under the pressure of four kilometers of seawater, that reduction in volume is greatly favored."

The reaction used by *Methanopyrus*, like nearly all other biological energy-supplying reactions, can be viewed as a river of electrons flowing downhill on an energy gradient. But if the flow is downhill, why don't the hydrogen and carbon dioxide simply react spontaneously to form methane, without waiting for help from *Methanopyrus*? The reason is the same as the reason that a bowl of sugar, sitting exposed to the air, doesn't burst into flames: there are energy barriers on the downward slope that prevent the reaction from occurring spontaneously at any reasonable speed. What *Methanopyrus* provides are enzymes—catalysts that act as tunnels through the energy barriers, like the tunnels that convey water through a dam. And by putting molecular "turbines" in these tunnels, *Methanopyrus* can use part of the liberated energy for what it really wants to accomplish: the synthesis of organic compounds and ATP. That, in a nutshell, is metabolism.

There's only one other major trick that our Life has dreamed up, and that is photosynthesis—in particular the advanced, oxygen-liberating photosynthesis that is used by cyanobacteria and green plants. These organisms take the low-energy electrons that emerge at the bottom end of the river and, using the light energy captured by chlorophyll, boost the electrons back up to the top again. It's just like those recirculating cascades that some people have in their yards: the electrons do work on the way down and get work done to them on the way back up. Photosynthesis has two major advantages over the lifestyle of *Methanopyrus* and similar organisms. First, it taps directly into a huge energy supply, far greater than the geothermal energy supply available at the deep-sea vents and elsewhere. Second, it dispenses with the need for minerals or gases as electron donors and acceptors. Photosynthesis gives you the freedom to go where you want, so long as you stay in the sunlight.

The realm of the deep-sea-vent colonies and the realm of photosynthesis seem like totally independent systems, each invested in its own way of making a living. But, in fact, they are not independent. Many of the organisms in the vent colonies, especially those that live in the cooler zones a little distance away from the emerging fluid, use the oxygen that is dissolved in the seawater—oxygen that was originally liberated by photosynthesis at the Earth's surface. Such organisms include not just the visible creatures of the colonies—the tube worms and crabs—but also some of the microorganisms. There are many microorganisms, for example, that make a living by transferring electrons from reduced forms of sulfur to oxygen. Without photosynthesis on the surface, none of these organisms could exist, and the vent communities would be much less diverse. That's part of the reason that Chris McKay is a "sunlight chauvinist": photosynthesis supports not only the organisms who practice it and those who eat them, but also many of the seemingly independent organisms that live by carrying out inorganic reactions.

Another group of creatures whose lifestyle seems quite exotic to us consists of the microorganisms that exist beneath the surface of the Earth. Not a few millimeters down, like the organisms at Badwater, but hundreds or thousands of meters down, far beyond the reach of the Sun's rays. Almost everything about these organisms is controversial, including in many cases the most basic question of all: Are they alive or dead?

According to microbiologist David Bulkwell, of Florida State University, more than nine thousand species of microbes have been found in the depths of the Earth, mostly by the examination of drill cores or water samples from deep wells. The majority of the finds have been in sedimentary rocks, which have pore spaces through which water can circulate. Some of the microbes live off organic material that was deposited with the sediments when they were originally laid down or that percolated down into the rock at a later time. Other microbes carry out profitable electron-transfer reactions on chemicals available in their environment—they may reduce oxidized sulfur (sulfate) to sulfide, for example. Because the temperature rises with increasing depth, many of the subsurface microbes are thermophiles or hyperthermophiles. If Stetter is right that the upper temperature limit for life is around 150 to 160°C, organisms might survive down to about 5 kilometers below the surface in continental crust, and perhaps to twice that in oceanic crust, which heats up more slowly with depth. But no one has actually found life that far beneath the surface.

Microorganisms are not only found in sedimentary rock. Todd Stevens and James McKinley, of Battelle, Pacific Northwest National Laboratory (the research institute associated with the Hanford nuclear facility in southern Washington State), have found microbial communities in water extracted from the Columbia River basalts.[8] These are layers of volcanic rock, several kilometers thick in total, that were formed during massive eruptions about 10 million years ago. Reactions between water and iron-bearing minerals in the rock produce hydrogen, and the microorganisms transfer electrons from the hydrogen to carbon dioxide, forming methane—the same reaction used by *Methanopyrus* at the deep-sea vents. The methane-producing organisms, which are the primary producers, supply organic compounds for a variety of other anaerobic microorganisms that share the same habitat. Thus, unlike the deep-sea-vent communities, the basalt ecosystems are entirely independent of solar energy.

These buried microbial communities can remain isolated from the surface ecosystems for extraordinarily long periods of time. Some microbiologists believe that the microbes became enclosed in the rock when it first formed. Since some of the rocks in which organisms have been found date from the Cretaceous period, this would mean that the microbes have been there for over 65 million years. As the rock becomes more deeply buried and compressed, the size of the water-filled pores within it decreases, until eventually the microbes may be trapped in solitary confinement. This kind of existence has both advantages and disadvantages. On the downside, there may be nothing left for the microbe to eat, but on the upside, there are no other organisms around to be eaten by. Therefore, it's been suggested, the microbes go into a state of "starving dormancy"—a near-death experience that can last virtually forever. The bulk of the Earth's biomass, some people have suggested, consists of these inert but living microbes, trapped in rock pores far beneath our feet.

Chris McKay is skeptical. "There's two reasons why it might not work," he tells us. "One is random thermal decay. If you're frozen in liquid nitrogen, or if you're in the permafrost on Mars, you might be okay; but at room temperature or above, things break down. Left-handed amino acids—the kind life uses—turn into right-handed amino acids. That's a disaster, and there's no way around it. And the other thing is radiation from the rock—from the uranium and thorium and potassium that's there. Over millions of years it adds up to megarads. If you're not working to repair that damage, you're dead."

Whether they are dead or dormant, the apparent ability of microbes to remain intact inside rock over geological periods of time has an important bearing on the search for life on other planets, as McKay is the first to acknowledge. We'll have more to say about this aspect of the matter in Chapter 3.

There's another reason to be interested in hyperthermophiles and other exotic microorganisms, besides what they have to tell us about possible habitats for life. They also can give us important clues about the course of early evolution on Earth and about the nature of evolution itself.

Probably the most familiar image in evolution is the "tree of life," in which all currently living species, as well as extinct species that were evolutionary dead ends, are represented as leaves. The chains of extinct organisms that led to the "leaves" are represented by the trunk and by branches of diminishing thickness. According to the nineteenth-century German biologist Ernst Haeckel, the trunk of the tree of life had three major branches: animals, plants, and single-celled organisms. In the 1930s, a Frenchman, Edouard Chatton, proposed that there was an even more fundamental branching of the tree, which gave rise to cells with and without recognizable nuclei. The organisms without nuclei were called "prokaryotes" (or "bacteria"), while the organisms with nuclei (including animals, plants, fungi, and single-celled nucleated organisms like amoebas) were called "eukaryotes".

Until the advent of molecular biology in the 1950s, the classification of organisms (a science called "taxonomy") depended almost entirely on morphological criteria: creatures were assumed to be related if they looked alike. Since prokaryotes don't have many interesting anatomical features, taxonomists tended to ignore them. But that changed with the development of techniques for sequencing DNA and RNA, which allowed organisms to be classified on the basis of their genetic similarities. A leader in this field has been Carl Woese, a molecular biologist at the University of Illinois.[9]

Woese focused on a single kind of RNA known as 16S rRNA, which is a constituent part of the ribosome, the cell's protein-synthesizing fac-

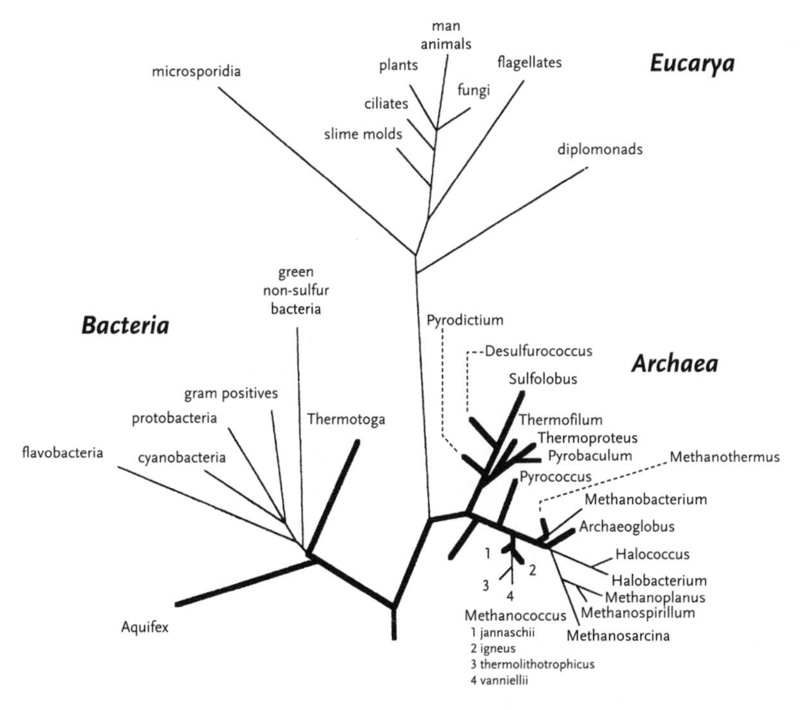

2.1 Carl Woese's 16s RRNA tree of life, modified by Karl Stetter to emphasize the parts of the tree occupied by hyperthermophiles (heavy lines). The position of the tree's root is speculative (courtesy of Karl Stetter).

tory. He chose it because it is abundant, it is present in all organisms, and it is a relatively short piece of RNA that can be sequenced without too much difficulty. Woese and his colleagues set out to sequence the 16s RRNA from organisms of many different kinds, from humans to the humblest prokaryote.

Unlike traditional taxonomy, which can make use of fossils, Woese's approach could only provide direct data about living organisms. But he used some fancy statistical footwork to reconstruct the tree's internal structure. Basically, Woese (or his computer) searched for a branching structure in which the distance between any two organisms, measured along the branches, was proportional to the degree of difference between their 16s RRNA sequences, and in which the total number of branches was minimized.

The resulting tree, it turns out, has three major branches rather than two: the previously recognized group consisting of eukaryotes forms one branch, but the prokaryotes fall into two groups, which Woese called "bacteria" and "archaea." Most of the familiar bugs—the *E. coli* that inhabits our large bowel, many disease-causing microorganisms, the cyanobacteria, and many others—are bacteria. The archaea are mostly rather unfamiliar organisms, many of which derive energy from exotic electron-transfer reactions involving minerals or gases. *Methanopyrus*, for example, is an archaeon. So too is *Halobacter*, the pink-colored organism in the surface layer at Badwater (the "bacter" in its name is a hangover from the days when all prokaryotes were considered bacteria).

Because RNA does not mutate at the same rate in all organisms, the 16S rRNA tree is not "time-labeled." Therefore, on the basis of the RNA data alone, Woese could not pick a single point on the tree and say, "This is the root, the spot from which all extant life originated." But he could make an estimate of the root's location by looking at genes that have duplicated and diverged from each other early in the history of life on Earth. Because these duplication events are one-way processes, they provided "arrows of time." Collectively they pointed back to a location on the tree, between the bacteria and the dividing point between the archaea and the eukaryotes, as being closest to the tree's root. Thus, the tree suggested a rather surprising conclusion: that we humans, as eukaryotes, are more closely related to the exotic archaea than to the humdrum bacteria.

Perhaps the most interesting thing about the tree, though, is that all the organisms nearest to the tree's root (in both the bacterial and archaean domains) are hyperthermophiles, and most of them are organisms that survive by catalyzing electron transfers among minerals or gases. Of all the currently living creatures that we know about, our lowly friend *Methanopyrus*, that lives in 100°C water and feeds off hydrogen and carbon dioxide, may be closest to the last common ancestor of all present-day life on Earth. These findings strongly suggest that the last common ancestor was also a hyperthermophile. We are all descended, it seems, from organisms that lived in near-boiling water!

Karl Stetter sees a connection to Günter Wächtershäuser's ideas about the origin of life, (described in Chapter 1). Life started at a high temperature near deep-sea vents, around 4 billion years ago, when volcanic activity was far greater than it is now and the Earth was still subject to frequent bombardment. It stayed hot for perhaps 200 million years, diversifying into hyperthermophilic bacteria, hyperthermophilic archaea, and possibly hyperthermophilic eukaryotes. If any nonhyperthermophiles evolved during this period, they were obliterated by the next major impact. Then, when the bombardment ceased, permanently cool environments offered themselves, and all three domains gave rise to organisms capable of living at what we call "normal" temperatures.

Not everyone agrees with this scenario. Leslie Orgel, for example, points out that a long period may have elapsed between the origin of life and the last common ancestor. (Attempts to date the last common ancestor by protein sequencing of modern organisms have yielded estimates ranging from 2 to over 3.5 billion years ago.[10]) Thus, even if our last common ancestor was a hyperthermophile, its ancestors might have lived at conventional temperatures. Adjustment to different temperatures, Orgel suggests, might be quite rapidly accomplished. Patrick Forterre, a French molecular biologist who has done much research on the molecular basis of heat tolerance, has proposed that the ancestral organisms were moderate thermophiles that lacked both gyrase and reverse-gyrase enzymes.[11] Some of these organisms, he suggests, evolved a reverse-gyrase enzyme and became hyperthermophiles, while others evolved a gyrase enzyme and became adapted to life at cool temperatures.

Furthermore, no hyperthermophilic eukaryote has yet been positively identified. Stetter has seen amoebalike creatures, almost certainly eukaryotes, on the hot surfaces of pieces of deep-sea vents brought up by submersibles. "It's not so easy to operate a microscope on a ship," he says. "Everything is trembling because of the diesel engine. But we saw them among the crystals of magnetite—we measured a temperature of about 100°C. Unfortunately, we have not been able to culture them so far. It's a tremendous challenge." If such organisms were confirmed to be eukaryotes, and to have 16S rRNA sequences that placed them near the base of the eukaryote line, the notion that we are all descended from hyperthermophiles would be greatly strengthened.

Carl Woese's tree of life, based on 16S rRNA sequences, is simple and elegant, but it may have significant shortcomings, as Woese himself has recently come to acknowledge. The major problem is that Woese's tree relies on analysis of a single gene, and it makes the assumption that this gene is representative of an organism's entire set of genes (or genome). This assumption turns out to be flawed.

We can see how misleading the single-gene approach can be if we think about disease-causing genes in humans. There are several rare genetic disorders, affecting perhaps a few hundred people in the United States, in which the defective gene is suspected of having been brought into the country by one affected individual, several generations ago. If we focus entirely on this gene, we can construct an evolutionary tree that links all the affected individuals to that immigrant, who becomes their "last common ancestor." But this ignores the fact that, thanks to sexual reproduction, each of us inherits one half of our genes from each of two parents, one quarter from each of four grandparents, and so on. Thus, if we select a different gene to study, that immigrant will almost certainly not be the common ancestor of all the affected individuals; in fact, he or she may not show up as an ancestor to any of them.

Thanks to the development of rapid gene-sequencing techniques, it has recently become possible to determine the sequences of many genes across different species. In fact, as of this writing, researchers have sequenced the complete genomes of fifteen organisms, including bacteria, archaea, and two eukaryotes (a yeast and a microscopic worm). It turns out that, indeed, evolutionary trees based on different genes are different from each other. If one focuses on genes that play a role in the copying of DNA into RNA and protein ("informational genes"), for example, one gets trees in which eukaryotes are more closely related to archaea than to bacteria, as in the 16S rRNA tree. If, however, one focuses on the much more numerous genes that code for enzymes involved in regular metabolism ("operational genes"), one gets trees in which eukaryotes are more closely related to bacteria.

There can only be one true history of evolution. Therefore, the metaphor of the "tree," while it may apply to single genes, seems not to adequately describe the evolution of organisms. Rather, we have to acknowledge that something analogous to sex happens, not just within species, but between species. Individual species must sometimes acquire their genetic endowment from more than one parent species. Instead of always diverging, the paths of evolution also sometimes rejoin, and the appropriate metaphor for evolutionary history may not be a "tree" but a "braided channel."

Transfer of genes between different species is, in fact, a well-documented phenomenon. Bacteria belonging to different species can pass genes for antibiotic resistance among each other, for example. One bacterium, *Agrobacter tumefaciens*, can insert foreign genes into the genomes of crop plants—it is much used in genetic engineering. And retroviruses such as HIV (human immunodeficiency virus) can insert genes into the genome of human cells (although not into germ-line cells). Thus, one way out of the paradox of the evolutionary data is to say that the "last common ancestor" was not a single species, but a large number of species that swapped genes among themselves.

Carl Woese has recently elaborated on this idea.[12] He has suggested that the first organisms were "progenotes"—organisms with very small genomes that replicated with many errors. These organisms readily swapped genetic material among themselves, so that the notion of discrete species was not yet a reality. Because of the high error rate, the small genome size, and the swapping of genetic material, evolution proceeded very fast at this stage. As evolution proceeded, genomes became larger, and the error rate decreased. These two processes went hand in hand because special proteins, requiring extra genes, were needed to control and perfect the replication process. Bacteria, archaea, and eukaryotes arose independently from this pool of progenotes, each domain helping itself, cafeteria style, from the genes available. In the process, the fidelity of DNA replication increased greatly, allowing much larger genomes, and the free exchange of DNA decreased, giving rise to discrete taxonomic groups, including species. Thus, there was never an individual "last common ancestor" that possessed all the ancestral genes of archaea, bacteria, and eukaryotes.

Evolution became more "treelike" after the separation of the three major domains, but genetic convergence still took place. One radical way in which this happened was by the fusion of entire individuals of different species. We already mentioned lichens—fungi containing

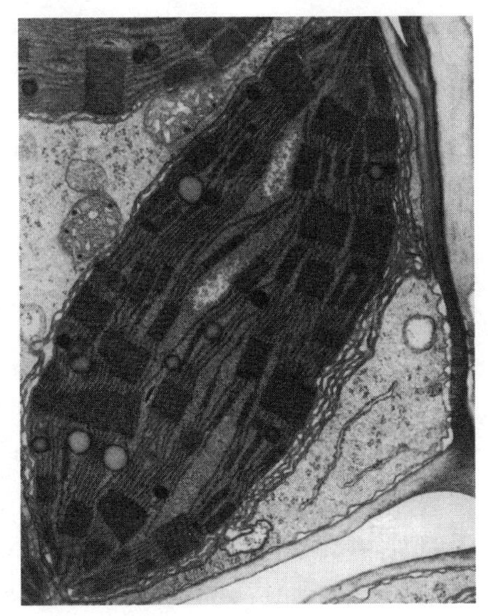

2.2 Ancient invaders? This is part of a leaf cell from a corn plant. The large oval structure occupying most of the field is a chloroplast; its complex internal membranes capture sunlight. Part of another chloroplast is seen at top left. Between the two chloroplasts are two smaller bodies containing dense particles and irregular membranes. These are energy-supplying mitochondria. Both chloroplasts and mitochondria are believed to have evolved from endosymbiotic microorganisms (electron micrograph courtesy of Lynn Margulis).

algae. This association is not a complete fusion into a single organism, however, because both the fungus and the alga each have their own genetic apparatus and are each capable of independent life. In addition, the algae are not topologically inside the cytoplasm of the fungus, but are enwrapped by fungal strands. Some symbiotic organisms do dwell entirely within the cytoplasm of their hosts, however; this phenomenon is termed "endosymbiosis."

According to Lynn Margulis of Boston University, endosymbiosis has, on many occasions, led to the complete integration of organisms with their host, so that the resulting hybrid cell became a single species.[13] The most clear examples, which are now universally accepted, concern the widespread cellular organelles called mitochondria and chloroplasts. Mitochondria are organelles within most eukaryotic cells that transfer electrons from simple organic compounds to oxygen, producing energy-rich ATP in the process. Chloroplasts are the chlorophyll-containing organelles within the cells of green plants that carry out oxygen-liberating photosynthesis. Both these types of organelles evolved from free-living prokaryotes that became endosymbionts within eukaryotic cells. Although most of their genes have been lost or transferred to the nuclei of their host cells, mitochondria and chloroplasts do still contain a few genes of their own, along with the machinery for translating them into RNA and protein. By sequencing these genes, researchers have identified relatives of mitochondria and chloroplasts among prokaryotes: mitochondria are related to the bacterium *Rickettsia prowazekii*, an intracellular parasite that causes epidemic louse-borne typhus, while chloroplasts have relatives among the photosynthetic cyanobacteria.

Margulis has suggested that several other eukaryotic organelles originated as endosymbionts but are not easily recognizable as such because they have lost all their genes to their host. Among them are the organelles called "flagella" (or, if there are many of them, "cilia"). These are motile, whiplike structures, projecting out from the cell surface, that either move the entire cell around (as with sperm) or move extracellular material past the cell (as with the ciliated epithelial cells of our trachea and bronchi, that move mucus out of the lungs).

Margulis believes that flagella originated as free-living bacteria called "spirochetes" (one species of which is the causative organism of syphilis). The great potential interest of this hypothesis is that it might explain how eukaryotic cells came by one of their key proteins, tubulin. This protein, along with another, actin, is responsible for the cytoskeleton of eukaryotic cells—the stiff but remarkably malleable scaffolding that gives these cells the ability to adopt an infinite variety of shapes and to engulf foodstuffs and other organisms. Microtubules, made of tubulin, form the core of flagella; so Margulis suggests that tubulin and actin, and perhaps the other cytoskeletal proteins, were originally a component of the endosymbiotic spirochetes.

This idea has not been widely accepted. First, there are no genes in the flagella themselves to sequence. And second, in spite of considerable work by Margulis and others, no one has been able to find a class of free-living organisms that has genes corresponding to the cytoskeletal genes in eukaryotes. In fact, the origin of these genes, so important to the evolution of eukaryotes, is still largely a mystery.

It's not known when exactly the ancestors of mitochondria entered their host eukaryotes. For a time, it was thought that it happened quite late in evolution, long after eukaryotes diverged from the bacteria and archaea, because some primitive single-celled eukaryotes, such as the intestinal pathogen *Giardia lamblia*, lack mitochondria altogether. But it's now clear that *Giardia* possesses several genes that are of mitochondrial origin: it must have got rid of its endosymbionts but kept some of their genes.[14] So it's possible that eukaryotes acquired endosymbiotic organisms almost as soon as the former diverged from archaea and bacteria.

The most daring hypotheses of this kind, however, attempts to explain the very origin of eukaryotes as a fusion event. One such hypothesis was put forward in 1998 by William Martin (of the Technical University of Braunschweig, Germany) and Miklós Müller (of the Rockefeller University in New York).[15] They propose that the first eukaryote was formed when an archaeon engulfed a bacterium. The bacterium, they say, consumed organic compounds and, in anaerobic conditions, released hydrogen and carbon dioxide. The archaeon was a completely anaerobic organism, like *Methanopyrus*, that consumed those same two gases and produced methane and organic compounds. Relationships between such organisms are obviously of mutual benefit; in fact, such relationships are the basis of the ecosystem in the Columbia River basalts, as described earlier. After fusion, the bacterial symbiont developed into mitochondria, but many of its genes were transferred to the host, which eventually gave up its hydrogen-consuming, methane-producing ways.

This hypothesis doesn't explain the origin of the eukaryote's cytoskeletal genes. A further wrinkle, however, has been added by Russ

Doolittle of the University of California, San Diego. He proposes that the fusion of archaeon and bacterium actually took place within the cytoplasm of a "protoeukaryote" that had previously engulfed them both, thanks to a cytoskeleton that the protoeukaryote already possessed. The resulting three-headed monster, Doolittle suggests, was so successful that it drove its simpler eukaryote siblings to extinction.

Whether any of these scenarios are correct is a question that will have to wait for the discovery and study of eukaryotes more primitive than any so far known. But the general message is clear: Evolution proceeds by divergence and by convergence of genetic information. All terrestrial life is linked, not just by some remote common ancestor, but by a process of genetic exchange and genetic merging that goes on today and will doubtless continue in the future. How could we humans, who have recently donated our genes to mice, pigs, and sheep in genetic-engineering experiments, fail to be aware of that fact?

Chris McKay can certainly relate to these ideas, for he is concerned with their possible validity on an even larger scale. Like Leslie Orgel, he takes very seriously the possibility that life can move from planet to planet in meteorites. He reminds us that, during the late, heavy bombardment, meteorites must have been exchanged between Earth and Mars much more commonly than they are today, and some of those meteorites may have contained living organisms. "Earth and Mars were swapping spit on a regular basis," he says. "Something very likely got through, perhaps many times in both directions."

Did Earth and Mars once form a single gene-swapping ecosystem? Only the direct study of martian life—if such exists or ever existed—can give the answer. But McKay has a strong preference. "If it turns out to be one Life, I'd be very disappointed. I'd be crushed. What I want to find is an independent origin of life—a second Genesis. That would be an intellectual watershed. Right-handed amino acids, say, instead of left-handed—that would be remarkable."

SUGGESTED READING

Koki Horikoshi, K., and W.D. Grant, eds. *Extremophiles: Microbial Life in Extreme Environments.* New York: Wiley-Liss, 1998.

Fredrickson, J.K., and T.C. Onstott. "Microbes Deep inside the Earth." *Scientific American,* October 1996, 68–73.

3

The Incredible Shrinking Martians

Searching for Life in the Solar System

Back in his office at NASA's Ames Research Center, Chris McKay seems a little uncomfortable. Although his office is bigger than those that many scientists enjoy, he's a bit too tall and rangy for the space available. His habit of leaping up and hiking over to his filing cabinets, to pull out papers and more papers on Mars and its possible inhabitants, accentuates the sense that he's yearning for the wide-open spaces.

The building that houses McKay's office is a nondescript structure on the perimeter of the research center. We find out later, in conversation with the 10-year-old son of a friend, that Ames's real attractions lie elsewhere: the wind tunnel so powerful that it can "blow your face to the back of your head" and the giant centrifuge that accomplishes the same trick with g forces. To hear him tell it, Ames is northern California's answer to Los Angeles's Magic Mountain Theme Park, with just enough science thrown in to make it a legitimate school outing.

But NASA is serious about its educational mission—and that, no doubt, explains why McKay is willing to sit down with us for a couple of hours. For NASA is anxious to be known for more than gung-ho rocketry, and it therefore expends a lot of effort in communicating its scientific programs, which include the study of life in the cosmos. Ames scientists have long played a major role in this enterprise. Their high point was the Viking mission to Mars in the late 1970s. While other Mars probes were exploding on the launch pad, crashing into the planet they were intended to study, or inexplicably breaking off radio contact, the twin Viking spacecraft performed flawlessly. While the orbiting motherships photographed the planet's surface at high resolution for more than a martian year, the landers touched down gently at their appointed spots and carried out an ingenious and highly successful set of experiments aimed at detecting life. Successful, that is, if you don't mind that they didn't detect any.

The twentieth century was a bad one for Martians, or at least for Earthlings' perception of them. A hundred years ago, Martians were intelligent, energetic creatures, capable of undertaking vast public works and even of launching expeditions against Earth. By the century's end, the Martians had been reduced to microbes, cowering in dark crannies far beneath the planet's frigid surface. They were most likely dead—and might even have never lived.

To properly recount the decline and fall of the Martian Empire would take the industry of a Gibbon. Here, we focus on a few highlights, hoping to illustrate the Martians' most salient characteristic—an uncanny ability to claw their way back into existence just when they seemed finally to have been obliterated.

Mars has long been known to resemble Earth in many ways. Like all the inner planets, it's made largely of solid rock. That fact alone makes it a more plausible candidate to harbor life than the four gas giants in the outer part of the solar system. The diameter of Mars is only 53 percent that of Earth (and martian gravity is only about 38 percent of terrestrial gravity), but because Earth's surface is two-thirds water, the land surface on the two planets is about the same. Mars is about 50 percent farther from the Sun than is the Earth; the average intensity of sunlight there is about 43 percent of what we experience on Earth, so a clear day on Mars is like a lightly overcast day here—plenty bright enough for photosynthesis.

The martian day is similar to the terrestrial day: 24 hours and 40 minutes. Like the Earth, the axis of Mars's rotation is tilted with respect to the plane of its orbit around the Sun, so Mars has seasons as we do. With the seasons, its two white polar caps, which resemble in appearance the Earth's ice-covered polar regions, shrink and expand considerably. The martian year is longer than ours, however, because Mars, being farther from the Sun, has a longer orbit to travel and moves more slowly in that orbit: its year lasts 687 Earth days. Like a faster runner on an inside track, the Earth keeps lapping Mars—that happens once every 2.14 terrestrial years. The moment when Earth passes Mars is called an opposition, because Mars and the Sun are then on opposite sides of the Earth.

Oppositions are good times to observe Mars, or to send rockets to Mars, because the two planets are then close to each other. But not all oppositions offer equally good opportunities. This is because (as discovered by Kepler) Mars's orbit is quite elliptical, while Earth's orbit is nearly a perfect circle. If Earth passes Mars when Mars is close to the Sun (which happens about every 15 years), the two planets come within about 0.38 of an astronomical unit of each other; but during oppositions when Mars is far from the Sun, they come no closer than about 0.67 of an astronomical unit. The eccentricity of Mars's orbit also has consequences for Mars itself: the seasons are asymmetrical, with southern summers being warmer but shorter than those in the north. Over long periods of time, however, this relationship changes.

In the Introduction, we already mentioned Percival Lowell, the American gentleman-astronomer who described the "canals" of Mars. The Lowells of Massachusetts were known equally for their wealth and their poetic imagination, and Percival was able to indulge both qualities to the full in his chosen field of study. He had a large observatory built at Flagstaff, Arizona, for the express purpose of finding life on Mars. "There is strong reason to believe that we are on the eve of pretty definite discovery in the matter," he told a Boston audience in May 1894, before setting out for Arizona to catch a favorable opposition.[1] By the following year, he had published a book laying out his discoveries: The "channels" previously described by Schiaparelli were artificial waterways, designed and built by Martians to bring irrigation water from the polar ice caps to the arid equatorial regions. The canals themselves were too fine to see, of course, but Lowell supposed that he was observing the strips of agricultural land irrigated by the canals, just as a Martian astronomer might spot the green Nile Valley but not the Nile River itself. He also saw "constructed oases" and other features. Lowell published two more books on the topic over the following 13 years and gave many public lectures. The presence of intelligent life on Mars, it seemed, had been scientifically established.

Already in 1894, however, critical voices were heard. E.W. Maunder, the British sunspot expert, suggested that the "canals" were optical illusions—the visual system's response to the presence of complex, unresolvable details in the image of the martian surface. But by then, Mars was fast receding, and it wasn't until the next favorable opposition, in 1909, that observational evidence contradicting Lowell's findings was obtained. This was the work of Eugene Antoniadi, a Greek-born astronomer working in France. He reported that, where Lowell had seen fine lines, there were only the margins of irregular blobs or slight differences in lightness. Antoniadi's report, although not universally accepted, marked the beginning of the end of the "canals" as a scientifically respectable topic. Ironically, Antoniadi was no enemy to the notion of intelligent life on Mars. "The presence of animals or even human beings on Mars is far from improbable," he wrote, but he suggested that such life might have died out for lack of water, leaving only plant life there today.[2]

Valid or not, Lowell's work seized the world's imagination and was the impetus for two seminal works of science fiction—one British, one German: H.G. Wells's *War of the Worlds*[3] and Kurd Lasswitz's *Auf Zwei Planeten* (Two planets).[4] Both books were published in 1897, and both concern Martian expeditions to Earth. Wells's Martians were monstrous in form and evil in intent: only their susceptibility to terrestrial microbes saved humanity. The German novelist, in contrast, portrayed Martians as resembling humans and as even being capable of interbreeding with humans; originally well intentioned, they were driven to repressive measures by rash human responses, such as the bombardment of a party of Martians by a blockheaded British naval commander. Parables both, the books inducted aliens into the service of moral discourse, a function they still perform admirably today.

Simply gazing through a telescope could not answer questions that were basic to the existence of life on Mars: What was the planet's atmosphere made of, and what was the temperature on the martian surface? In the early part of the twentieth century, other techniques were brought to bear on these questions, most notably spectroscopy.

Spectroscopy involves breaking up light into a rainbowlike spectrum of wavelengths with a prism or grating. Because many elements and compounds absorb light at specific wavelengths, the presence of these elements or compounds between the light source and the observer can be recognized on the basis of the characteristic dark bands they produce in the spectrum. When observing Mars, for example, a set of bands might be produced by a certain compound in the outer layers of the Sun, in the martian atmosphere, on the martian surface, in the Earth's atmosphere, or somewhere along the way. The correct location can in principle be determined by making comparisons between observations of different targets: for example, bands that show up when observing Mars but not when observing the Moon are likely due to elements present on Mars or in its atmosphere. Spectroscopy can often determine not only the existence but also the amount of a compound that is present.

Some early spectroscopic observations—for example those conducted by V.M. Slipher at the Lowell Observatory in 1907—suggested the presence of abundant water vapor and oxygen in the martian atmosphere, circumstances obviously very favorable to the existence of life. But subsequent studies came up with lower and lower estimates. By 1926, oxygen was down to "undetectable" levels, and water vapor was present at no more than 3 percent of its terrestrial abundance. In spite of these developments, belief in the existence of plant life on Mars persisted and even strengthened. In 1947, a Dutch-born American astronomer, Gerard Kuiper, correctly detected the telltale bands of carbon dioxide in the infrared region of Mars's spectrum. This raised the possibility of photosynthesis, since photosynthetic plants use carbon dioxide as their carbon source. Kuiper failed to detect chlorophyll, but he suggested that martian plant life might resemble terrestrial lichens, some of which use other kinds of molecules to capture light. Then, in 1957, William Sinton of Harvard University reported that he had detected absorption bands further into the infrared, which he (wrongly) ascribed to the organic components of plant life (including carbohydrates). Therefore, he concluded, there was an extensive cover of vegetation on Mars.

During the 1960s, three Mariner spacecraft flew by Mars and sent back pictures of a rugged, cratered terrain without any signs of life. They also sent back data about the atmosphere and surface that made clear just how hostile to life Mars really is. According to the latest measurements, the average atmospheric pressure at the surface is less than one hundredth the pressure on Earth. The atmosphere is 95 percent carbon dioxide and 3 percent nitrogen, with traces of other gases. There is almost no oxygen, and so little water vapor that, if it were all squeezed out, it would form a layer of liquid water only about one hundredth of a millimeter thick on the planet's surface. Even so, water-ice clouds do sometimes form. Mars is bone-chillingly cold: even at the equator, the average temperature is only about minus 60°C, and the warmest it ever gets is about 7°C. At the polar caps, which consist of solid carbon dioxide and an underlying layer of water ice, temperatures remain at minus 123°C all winter, rising to a balmy minus 80°C at midsummer. Because of the combination of low temperature and low pressure, liquid water cannot exist on the surface: it would either freeze or boil. The early Mariner flybys served up a harsh dose of reality for those who believed that Mars possessed a biosphere resembling Earth's.

In 1971, Mariner 9 went into orbit around Mars and photographed the entire surface of the planet. Although it did not find any more signs of life than its predecessors did, it revealed many interesting features that had escaped earlier observation: great volcanoes, deep canyons, and what looked like ancient river valleys.

Having this new map of Mars in hand, Carl Sagan and Paul Fox, of Cornell University, reexamined Percival Lowell's canal hypothesis.[5] They overlaid the Mariner map with Lowell's canal map, looking for corresponding features. They found essentially none. Thus, the "canals" seem to have been entirely the products of the eyes and brain of Lowell and the other astronomers who claimed to see them.

Still, the discovery of what seemed to be canyons and river valleys sparked the idea that Mars was once warmer and wetter than it is now. If so, maybe life evolved there in the past and perhaps was able gradually to adapt to the changing conditions or had left spores that could be reawakened by warmth and moisture.

This, then, was the thinking that guided the Viking biology experiments of 1976–77. Each of the landers carried a "biology package," about the size of a small television set, which contained equipment to carry out three different experiments: a "pyrolytic-release" experiment, a "gas-exchange" experiment, and a "labeled-release" experiment, each designed by a different research team with a different philosophy about how to detect life.

In the pyrolytic-release experiment, developed by a team led by Norman Horowitz of Caltech, the idea was to alter the environmental conditions as little as possible. Radioactively tagged carbon dioxide and carbon monoxide, but no water, were added to samples of martian soil. The sample was left to incubate under a sunlamp for a few days, so that any organisms in the soil would have a chance to incorporate the radioactive gases into organic compounds, using photosynthesis if they so wished. Then the soil was heated to 635°C ("pyrolyzed") to vaporize any organic compounds that had been produced, and these were measured with a radiation detector. Although the early results of this experiment were positive, there turned out to be no consistent difference between the results obtained with untreated soil and those obtained with soil that had previously been heat sterilized—a crucial control. Thus, Horowitz eventually concluded that the experiment offered no evidence for the presence of any living organisms in the soil.

The gas-exchange experiment, developed by Vance Oyama of Ames, involved adding water vapor, and then organic compounds, to a soil sample. The idea was that any microbes in the soil would be reawakened into metabolic activity and would produce gases that were to be detected and identified by a gas chromatograph. Two gases, oxygen and carbon dioxide, were indeed given off by the soil samples, but from the rate and time course of the reaction, Oyama concluded that the release was due to the presence, not of living organisms, but of a powerful oxidant—perhaps hydrogen peroxide—in the soil.

In the labeled-release experiment, designed by Gilbert Levin, a solution of radioactively labeled compounds in water was added to a soil sample. The sample was kept warm and wet, mimicking conditions that most terrestrial microbes like. Any gaseous products given off were measured in a radiation detector. There was in fact a surge of gas given off after the nutrient mixture was added to the sample, followed by a slower, steady release. Presterilizing the soil eliminated the release. Thus, this experiment, considered by itself, did offer some evidence in support of the existence of living organisms in the martian soil, and Levin held to this interpretation for years after the mission was over.

In spite of this finding, another Viking experiment (not part of the "biology package") helped deflate the hopes for life. A gas chromatograph/mass spectrometer, designed by a team at MIT, searched the soil for the presence of organic compounds—and found none. Given the extreme sensitivity of the instrument, the negative result was surprising. After all, organic compounds are known to be present in micrometeorites, as mentioned in Chapter 1, and so they should have steadily accumulated in the martian soil. There must be something that destroys organics as fast as they arrive; that something may be the unidentified oxidizing agent that played havoc with the gas-exchange experiment.

Thus, while some scientists such as Levin held on to an optimistic view of the matter, the general feeling was that the Viking landers had disproved the existence of life at the locations they examined. And because the wind on Mars tends to blow surface dust around the planet, it was hard to see how life could exist anywhere on the surface without producing some kind of a positive result at the two landing sites. There were suggestions that living creatures still might be found in some unexplored niches, such as the polar caps or possible hydrothermal springs, but the general effect of the Viking mission was to bring the notion of life on Mars to a low point of credibility.

Curiously, however, the Viking mission also held the seeds of a rebirth. While the landers were conducting the biology experiments, the two orbiters carried out a detailed photographic survey of Mars. It was the results of this survey, combined with the earlier Mariner data, that opened a new debate—one that is still raging—about the possible existence of life on Mars in some past epoch, when the planet was warmer and wetter than it is today.

"Basically, it boils down to a very simple argument," says McKay. "All the conditions on Earth that have been postulated to be critical to the origin of life have also existed on Mars. Hydrothermal systems, organics coming from comets, sunlight for photosynthesis, sulfur-rich CO_2 [carbon dioxide] chemical systems like Wächtershäuser looks at. Whatever theory you like, I'll make it work on Mars. Water? Yes, 3.8 billion years ago, when life had already appeared on Earth, water was flowing on Mars."

McKay brings out a picture that looks, at first glance, like a photograph of Earth as seen from the Moon: a sphere hanging in the blackness of deep space, with a ruddy continent set off by clouds and by the engirdling blue of a vast ocean. "I used this in an article for *Astronomy* magazine," he says. "It's a real case of a picture worth a thousand words. I show this to a general audience and say 'Mars may have looked like this once. This is not just a fantasy.' "[6]

To make his ocean, McKay relied heavily on the work of Michael Carr, a planetary geologist at the United States Geological Survey at Menlo Park, a few miles north of Moffett Field.[7] Carr has made a close study of what appear to be water-carved features on Mars's surface. The most spectacular of these features are the "outflow channels," many of which are found in the region around the Chryse Basin in Mars's northern hemisphere, just to the east of the towering volcanoes of the Tharsis Highlands.

The outflow channels were formed when highly pressurized subter-

3.1 The remains of a great martian flood. The triangular depressed area is named Ravi Vallis. The irregular blocks in the upper right portion of the area form "chaotic terrain," where the land surface collapsed after the release of huge quantities of subterranean water. The floodwaters ran downslope (to the lower left) scouring out a deep channel. The scene is about 250 kilometers wide (Viking Orbiter image, courtesy of Tim Parker, JPL).

ranean water burst through a confining lid of permafrost and erupted as catastrophic floods onto the planet's surface. The regions where the eruptions took place are known as "chaotic terrains": chaotic because the land surface collapsed in irregular blocks after the underlying water was ejected. The channels scoured by the raging floodwaters lead from the chaotic terrains downhill to the northern lowlands. The size of some of the channels is truly impressive: at the peak of each outflow, which probably lasted no more than a few days or weeks, water must have flowed at rates of up to a billion cubic meters per second—the equivalent of ten thousand Mississippi rivers in full spring flood. Each flood delivered up to 300,000 cubic kilometers of water to the northern lowlands, and the combined total of water delivered by all the floods (estimated from the amount of material eroded away to form the channels) was at least 6,000,000 cubic kilometers. This volume of water, if it were spread evenly over the entire planet, would form a layer 40 meters thick.

But there was much more water than that. The regions that were subjected to the catastrophic floods form only about 10 percent of Mars's surface. Carr believes that there was just as much subsurface water in other regions of the planet, only it never broke through to the surface. Thus, he estimates the total inventory of water on Mars, early in its history, as being equivalent to a 400-meter-thick layer spread over the whole planet. (For comparison, Earth's total inventory of water, spread evenly over the Earth's surface, would form a layer about 3 kilometers thick.)

McKay created his ocean simply by taking this 400 meters of water and pouring it onto Mars. It collected in the low-lying northern plains, forming an "Oceanus Borealis," or northern ocean.

Carr, however, is as miserly with "his" water as McKay is spendthrift. When, on another occasion, we talk with Carr directly, he tells us what he thinks is wrong with the concept of a martian ocean. First, he says, most of that 400 meters of water never broke through onto the surface and is still locked deep under the permafrost. Second, the flood events probably didn't happen all at once but were well scattered in time. What really happened, Carr thinks, is that the water from each flood formed a lake. Some of the water in the lake may have soaked back into the subsurface, and what remained on the surface rapidly froze solid and has stayed frozen up to the present. No body of liquid water remained exposed on Mars's surface for geologically significant periods of time.

McKay, though, is by no means alone in his point of view. Vic Baker, of the University of Arizona, has been a long-time proponent of martian oceans; in fact, it was he and his colleagues who came up with the name Oceanus Borealis.[8] Baker believes that Mars has had episodes of substantially warmer temperatures than it now enjoys. These warming episodes may have been linked to long-term changes in the planet's orbital characteristics and in the orientation of its axis of rotation. Such warming might have triggered flood events by thinning the permafrost cover. In addition, however, the flood events would contribute to the warming, by virtue of the water vapor and carbon dioxide—both greenhouse gases—that would enter the atmosphere (the carbon dioxide, derived from carbonate rocks, would presumably have been dissolved in the trapped groundwater and released along with it).

The warm, wet spells, in Baker's view, were not confined to the early "Noachian" age of Mars's history (the time of the late bombardment, which is thought to have ended about 3.8 billion years ago) but recurred in the ensuing "Hesperian" and "Amazonian" ages. These two periods are defined by cratering densities and have not been given definitive dates. Still, the thinking of Baker and his colleagues is that the most recent warm episode may have been no more than 300 million years ago.

If a liquid ocean had been pumping water vapor into the martian atmosphere, there must have been a water cycle; that is, the water vapor must have precipitated out of the atmosphere as rain or snow and returned to the ocean in rivers or glaciers. There are in fact many features on Mars that have been interpreted as branching river valleys and as the beds of glaciers. The majority of them are found on the ancient, heavily cratered terrain of the Noachian period. Here too, however, there is considerable controversy. Carr suggests that many of the "river valleys" did not collect rainwater but were formed and fed by springs (a process termed "groundwater sapping"). Many other channels, he thinks, were formed by "mass wasting," the mass downhill movement of jumbled rock lubricated by small amounts of water. The minority of true river valleys, he thinks, belong to the Noachian period and therefore don't support the idea of wet epochs late in Mars's history.

Carr has another line of argument against the late wet periods. This has to do with erosion rates. On Earth, even the most arid regions lose between 0.01 and 1.0 millimeter (mm) of their surface per year to erosion. Study of crater rims and other features on Mars has shown that, during the Noachian age, the planet eroded at a rate corresponding to the bottom end of that range (about 0.01 mm per year). This is consistent with the idea that some precipitation did occur during the Noachian age. Since the end of the Noachian age, however, the average erosion rate has been a thousand times slower. Thus, if there were warm, wet periods late in Mars' history, they must have been extremely brief.

Another player in the ocean controversy has been Tim Parker of Caltech's Jet Propulsion Laboratory (JPL).[9] While involved in the MX missile program in the 1980s, Parker did research on the Great Basin area of Nevada and Utah. The Great Basin once held several enormous lakes, which are now reduced to salt deserts. The shorelines of these ancient lakes can still be traced as wave-cut platforms around the perimeter of the salt flats. So after turning his attention to Mars, Parker wondered whether oceans might have left shoreline features there too, and he pored over the Viking and Mariner photographs in search of them.

He found what he considers evidence for not one, but nine concentric shorelines, each of which marks the perimeter of the Oceanus

Borealis when it was at a different size. Some he has traced over thousands of kilometers, while for others he has only found fragmentary indications. The most recently discovered, outermost shoreline, which Parker studied in collaboration with Ken Edgett (now of Malin Space Science Systems in San Diego), was the earliest of the oceans and may have formed by condensation of water out of Mars's steamy primordial atmosphere. The ocean covered about half the planet; and if its water were spread out over the entire planet, it would have been 1 kilometer deep. Thus, it held much more water than Carr's estimate of Mars's total water inventory. Because this amount of water is more than could be hidden beneath the surface of present-day Mars, Parker believes that much of the water was subsequently lost to space: through photodissociation and subsequent escape of hydrogen, through erosion by the solar wind, or through bulk loss of atmosphere during major impacts.

Carr has definite reservations about Parker's work. "The problem is that we don't have very good photographic coverage of these high northern latitudes where these oceans are most likely," he told us. "There are a lot of strange linear features that one sees. It would be interesting to see if they connect. Unfortunately, we can only see these features in very high-resolution images, which show only a tiny area, so we don't have much sense of whether they are continuous." Ultimately, the nature of these "shoreline features" will be resolved by more-extensive mapping, such as the photographic and altimetric (height-measuring) survey now being undertaken by the Mars Global Surveyor spacecraft.

Carr, Parker, and Baker are primarily geomorphologists: they study landforms. But atmospheric scientists have also taken part in the debate about Mars's history. The general feeling among these scientists is that water vapor and carbon dioxide by themselves would not have been sufficient to warm Mars's surface to above the freezing point of water, particularly in an era when the Sun is thought to have been only 70 percent as bright as it is now.

One possible solution is that of Sagan and Chyba, already discussed in Chapter 1 in the context of Earth's early climate: This is the idea that more-potent greenhouse gases, such as methane, might have been present in the atmosphere and protected from destruction at the hands of the Sun's ultraviolet radiation by a shielding layer of "smog." More recently, François Forget (of the Université Pierre et Marie Curie in Paris) and Raymond Pierrehumbert (of the University of

Chicago) have suggested that a strong greenhouse effect was exerted by carbon dioxide, not in gaseous form but as clouds of dry-ice particles high in the atmosphere.[10] Forget and Pierrehumbert's calculations indicate that such clouds would prevent the loss of infrared radiation to space, not by absorption (which is the way greenhouse gases work), but by reflection of the radiation back to the surface. This mechanism, it seems, might also have played a role in keeping Earth warm in its early days.

In spite of the controversy about the age and extent of oceans on Mars, there does seem to be broad agreement that there was some liquid water on Mars's surface around 3.8 billion years ago, as the late bombardment was coming to a close. And to McKay, the message seems clear. "If you look at Earth," he says, "in some sense that's the defining ecological parameter for life—liquid water. It's not the only thing life needs, of course. But of all the requirements for life, water is the only one that's rare."

Showing that conditions suitable for life once existed on Mars doesn't mean that life actually arose there, of course. The question of early life on Mars would have remained entirely hypothetical had it not been for the dramatic announcement, in August 1996, of evidence for a relic of life in a martian meteorite. This study was the work of a team led by geologist David McKay (no relation to Chris), who is at NASA's Johnson Space Center in Houston.[11]

The chunk of rock in question belongs to the very rare SNC class of meteorites—only fourteen examples have been found. The name is made up of the initials of three of the locations where the meteorites fell: Shergotty, India (in 1865); El Nakhla, Egypt (in 1911); and Chassigny, France (in 1815). The SNC meteorite studied by McKay's group, however, was not observed to fall but was simply found lying on the ice in the Allan Hills area of Antarctica in 1984: it has been given the unromantic name ALH84001. It had lain buried in the ice for several thousand years, before it was exposed by the scouring action of the wind.

One thing about which everyone agrees is that the SNC meteorites do indeed come from Mars. They clearly don't come from Earth, because the relative abundance of different isotopes of oxygen in them is quite different from what is found on Earth. But the chemical composition of the gas trapped in tiny bubbles within the meteorites, as well as the proportions of different isotopes, closely matches the measurements of the martian atmosphere made by the Viking landers.

ALH84001 is a bit of an odd-man-out among the SNC meteorites. Although, like the others, it formed by solidification of volcanic magma, it is by far the oldest: radiometric measurement of the decay of several elements in the meteorite indicates that the rock solidified 4.5 billion years ago—only about 100 million years after Mars first formed. The other SNC meteorites formed between 1.3 billion and 150 million years ago (long after the epoch when Mars seems to have been most hospitable to life).

The history of ALH84001, since its original formation 4.5 billion years ago, has been deduced from examination of its internal structure. Shock-induced fractures in the rock were probably created during the late heavy bombardment of Mars, around 3.9 to 3.6 billion years ago, while ALH84001 was still part of its parent rock formation. In some of these fractures, McKay's team found globules or pancakelike discs of carbonate—mineral forms of carbon probably derived from the interaction of carbon dioxide and salts dissolved in water. The globules are up to about a quarter of a millimeter in size. The presence of these globules indicates that water percolated through the fractures in the rock at some time after the impacts that caused the fractures.

At least one more impact pounded the rock after the late heavy bombardment was over. The evidence for this is that some of the carbonate globules sitting in the early fracture zones are themselves split by another set of fractures that cuts across the first. It was yet a third impact, a mere 14 million years ago, that sent ALH84001 on its journey to Earth. That date is known from measurements of isotopes of the noble gases—helium, neon, and argon—which gradually build up in meteorites in space as they are irradiated with cosmic rays. Another isotope that builds up in space, carbon 14, decays gradually to nitrogen 14 after the meteorite leaves the space environment. By measuring how much of this carbon 14 was left in the volcanic minerals of the meteorite, Timothy Jull, of the University of Arizona, was able to determine that the meteorite fell to Earth about 13,000 years ago.

It doesn't necessarily take 14 million years for a piece of Mars to reach Earth, by the way. One of the SNC meteorites left Mars only 0.6 million years ago. And according to modeling by Brett Gladman of Cornell University, the impacts that kicked the SNC meteorites into space probably sent some rocks on trajectories that intercepted Earth within less than a year.[12] This conclusion, if correct, makes it a lot easier to believe that living organisms could survive the journey, even without a great deal of shielding.

McKay's team does not claim to have found a "clincher"—a single piece of evidence that would be irrefutable proof of the past existence of living organisms on Mars. Rather, the team points to a number of features of the meteorite, mostly to do with the carbonate globules, that the team believes can collectively best be explained by biological processes.

First, within the carbonate globules, the researchers found minute crystals of magnetite (Fe_3O_4). These crystals, measuring only about 50 nanometers across (a nanometer is 10^{-9} meters), are similar to those found in many terrestrial bacteria, where they serve as little compasses. By orienting themselves with respect to the slanting lines of the Earth's magnetic field, the bacteria are able to tell "up" from "down."

Second, the researchers found organic carbon compounds in association with the globules. These compounds included a class of substances called polycyclic aromatic hydrocarbons, or PAHS, that the researchers interpreted as being the decay products of microbes that became trapped in the forming globules.

Third, and most dramatically, McKay believes that he and his colleagues have seen the martian microbes' fossilized remains. Using very high-resolution scanning electron microscopy, McKay and his colleagues found areas within the carbonate globules that contain ovoid or tubular structures resembling the tiniest bacteria found on Earth. A photograph of one particularly striking example—a segmented, wormlike object resting languidly on a bed of mineral grains—has appeared in newspapers, in magazines, and on television shows around the world. This 'creature'—if 'creature' it is—has become Mars's de facto ambassador to Earth.

McKay's announcement ignited a firestorm of controversy. Some critics simply asserted their reasons for thinking that McKay's conclusion had to be wrong. It was pointed out, for example, that Mars lacks a magnetic field, so martian bacteria would have no use for magnetite particles. It was also claimed that the carbonate globules must have formed at temperatures far too high for living things to have existed inside them. Many other scientists dropped whatever they were doing, beseeched NASA for a chip of the meteorite, and conducted their own tests. Among the latter was Jeff Bada, the San Diego origin-of-lifer whom we met in Chapter 1.

Bada (with several colleagues) studied the organic materials in the meteorite.[13] He looked particularly for amino acids in the carbonate globules. He found them, but their kinds and relative concentrations resembled what is found in Antarctic ice. So Bada summoned a press conference and denounced the organic compounds in the meteorite as terrestrial contaminants. "I've been saying this all along," he tells us during our visit to San Diego, "and it's proven to be the case. It's bad news for the 'pro-lifers.' And the kiss of death is, if you do a step pyrolysis [gradual heating] of the meteorite and collect the CO_2 [carbon dioxide], it's got carbon 14 in it. There's no way you can have carbon 14 in a 4-billion-year-old meteorite [because it would have all decayed back to nitrogen]."

The carbon 14 study was actually done by another group (led by Timothy Jull), and the results were not quite as damning as Bada would have us believe.[14] First, some carbon 14 was produced in the meteorite while it was in space, as described earlier. More significantly, Jull's group actually found that one fraction of the carbon dioxide released during heating contained no carbon 14 at all. This fraction must be of extraterrestrial origin. It is not from the carbon minerals within the globules, because the researchers dissolved away the carbonate with acid prior to their analysis. Thus, it could be from organic compounds produced by the martian "microbes" when they were alive.

Another person who got a piece of ALH84001 was Joe Kirschvink of Caltech.[15] Kirschvink is an expert on geomagnetism and has applied his expertise to all kinds of topics: for example, he has measured ancient movements of the San Andreas fault by studying shifts in the magnetism of the ground. Kirschvink knew that terrestrial volcanic rocks, when they solidify, become weakly magnetized. This happens because crystals of iron-containing minerals align themselves with the Earth's magnetic field while the magma they are floating in is still liquid, but once the magma solidifies the crystals get locked in place. Kirschvink and his colleagues found that ALH84001 became magnetized when it solidified, and to about the same field strength as that of terrestrial rocks. Therefore, Mars, though it has no magnetic field today, did have a strong field early in its history. This conclusion has been bolstered by results from the Mars Global Surveyor spacecraft, which has detected magnetic anomalies in some regions of Mars.

These findings not only give martian microbes a reason to possess magnetite crystals; they also mean that early Mars had a magnetosphere, which would have protected the planet from the charged particles streaming from the Sun (the solar wind). Thus, the erosion of the

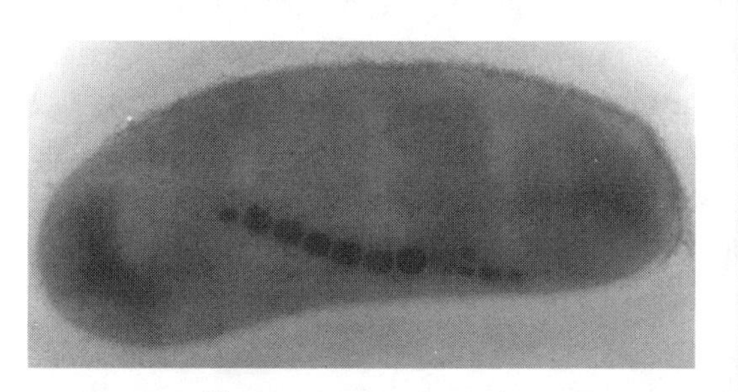

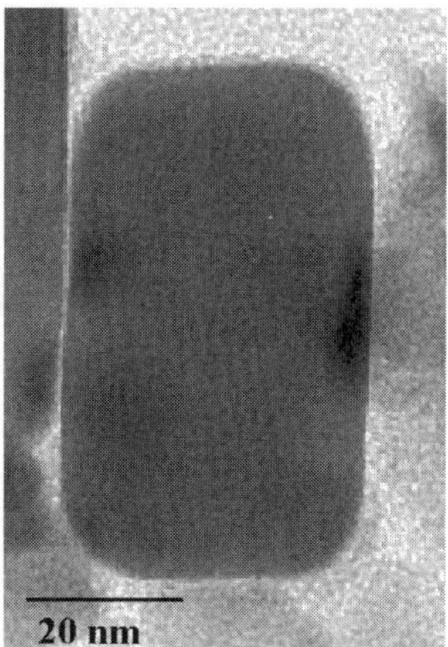

3.2 a) Terrestrial bacterium containing a chain of magnetite crystals (courtesy of Joseph Kirschvink).
b) Magnetite crystal (at higher magnification) from the martian meteorite ALH 84001 (courtesy of Kathie Thomas-Keprta).

atmosphere by the solar wind, which contributed to the cooling and drying of Mars, was postponed until a later epoch.

The locked-in, or remanent, magnetism of a rock sample is lost if it is heated above a certain critical temperature, which depends on the kind of magnetic minerals that the rock contains. In a feat of sleuthing that Sherlock Holmes himself might have been proud of, Kirschvink noticed that a few tiny fragments of the meteorite had been rotated through a small angle by the shock that created the first set of fractures. The orientation of the remanent magnetic fields within these fragments rotated with them, and has remained fixed ever since. Kirschvink concluded that ALH84001 could not have been heated much above 110°C during the time when McKay's postulated microbes were living, otherwise the orientation of the magnetic fields in the rotated chips would have been at least partially erased and reset to the ambient orientation. In other words, it was never too hot for a self-respecting hyperthermophile to flourish.

McKay's interpretation of the magnetite crystals as products of martian microorganisms took a hit in 1998, however. A group led by John Bradley (of MVA Inc. in Norcross, Georgia) studied the structure of the crystals at high magnification.[16] The group found that the axes of some of the crystals were perfectly aligned with those of the carbonate crystals that they were touching. The group's interpretation was that the magnetite crystals grew while in contact with the carbonate crystals (called "epitaxial growth"). This interpretation is inconsistent with the notion that magnetite crystals grew inside cells and were released into the environment after the cells died.

Bradley's conclusions are contested by Kirschvink, who tells us that the occasional alignment of the magnetite and carbonate crystals could have occurred by "epitaxial settling": in other words, the magnetite crystals, after the organisms that formed them died, simply fell onto the carbonate crystals and locked into their crystal lattice, like Velcro. Imre Friedmann, who has studied pieces of the meteorite by scanning electron microscopy, also strongly defends the biological origin of the magnetite crystals. He tells us that the elongated shape of the crystals, and their alignment in chains, are unique to magnetite produced by microorganisms.[17]

Bradley and colleagues have also challenged the images of the "microfossils."[18] They claimed that the wormlike forms are nothing but the regular edges of mineral layers, fattened out a bit by the metallic coating that has to be put on the specimen to make it visible in the scanning electron microscope. In response to such criticisms, McKay's group (in particular, the electron microscopist, Kathie Thomas-Keprta) originally denied that there is any confusion between mineral edges and their microfossils, and the group presented micrographs in which the bacterialike structures are much less regularly aligned than in their original data. More recently, however, members of McKay's group have expressed themselves much more cautiously about the images. In particular, they have conceded that many of the "organisms" are too small to be bacteria—they would simply lack the space to house the most basic machinery of life.[19]

Another group has come up with a different kind of evidence against the "microfossils." Derek Sears and Timothy Kral, of the University of Arkansas, examined other meteorites also found in Antarctica, which are known to have originated on the Moon rather than on Mars.[20] In these meteorites, Sears and Kral found bacterialike objects like those in the martian meteorite. Because the Moon is thought to have been lifeless throughout its history, the researchers concluded that the microbelike structures are not fossilized extraterrestrial organisms but instead were formed during the meteorites' residence in Antarctica.

The idea that intriguing structures can form in meteorites after they land on Earth has received further support from the study of a (nonmartian) meteorite that fell in the Sahara Desert in 1931. A group of French researchers, led by J.A. Barrat of the University of Angers, compared samples of the meteorite collected right after it landed with samples collected 63 years later.[21] The researchers found that fracture surfaces in the original samples were "clean," but similar surfaces from the weathered samples displayed carbonate "pancakes" as well as tiny rods and spheres resembling the "microfossils" in ALH84001. Even in the arid Sahara, apparently, occasional wetting can lead to the deposition of carbonates and microbelike structures.

All in all, the hypothesis of David McKay's group, that ALH84001 was once home to martian microbes, seems quite a bit less compelling than when it was first put forward. The only piece of evidence that still has strong adherents is the magnetite crystals. When we ask Chris McKay about his namesake's findings, he is noncommittal. "I'm pro-choice rather than pro-life," he says. "Their evidence is consistent with life, but it's consistent with a lot of other processes too. They haven't proven their case, and their critics haven't proven them wrong either." To Chris McKay, the main significance of the martian meteorite is to highlight the importance of continuing the program of exploration begun by the Viking landers.

"If there's life on Mars today, it's probably deep underground," he says. "There could be organisms that live off CO_2 [carbon dioxide] and hydrogen, like the ones in the Columbia River basalts. There could be an ecosystem built around them. But to find them, you'd have to drill down through the permafrost—kilometers, probably. It would be a tough job. Maybe you could use ground-penetrating radar to find places where liquid water comes close to the surface. You might even be able to find a little vent, a geyser, where it comes all the way up to the surface: Mars's 'Old Faithful.' But that's unlikely."

McKay is involved in planning missions to test for the possibility of life on Mars, and his assessment is that it's not cost-effective to be looking for life that may currently exist beneath the martian surface. "My bias is towards finding fossilized organisms in ancient lake beds. Wherever water pooled on the surface, life may have existed in ice-covered lakes, as in the Antarctic lakes on Earth. Those are good places to look for fossils because the sedimentary materials would be there on the surface. And right now, we can point to places that were lake beds, like the Gusev Crater in the southern hemisphere. They're the most promising for a near-term mission."

Imre Friedmann makes a slightly different suggestion. He proposes that if life ever existed on Mars, it would have been driven into habitats like those of his Antarctic endolithic communities (see Chapter 2) as the planet cooled and dried out.[22] The trace fossils of these communities might still be detectable on rock surfaces.

McKay and most others in the field believe that the biological exploration of Mars will require the return of samples to Earth rather than analysis on Mars, and NASA has made plans accordingly. There's a mission planned for the year 2003 that will collect samples with a rover. The rover will deliver the samples to a lander, which will fire them into orbit. Another mission in 2006 will do the same thing somewhere else on Mars. Then a French-built sample return vehicle, launched in 2006, will rendezvous with one or both of the orbiting caches and return them to Earth by 2008.

We ask McKay why we should be going to Mars for rocks when Mars has already sent us a bunch. "All the Martian meteorites come from volcanic terrains," he says. "That's maybe because only volcanic rocks have enough integrity that when they're hit by a comet they stay intact and launch off into space. A block of sandstone would just turn to sand, and it will never make it. But volcanic rocks are not ideal for searching for fossils. If you wanted to find fossils on Earth, landing on Kilauea and taking samples would not be the ideal way. It would be better to land on the Bonneville salt flats and take a sedimentary core."

On Earth, 3.5-billion-year-old fossils, such as Bill Schopf has found, are extremely rare. But that's in large part because 3.5-billion-year-old rocks are themselves rare, thanks to the constant recycling of the Earth's lithosphere. In addition, when such rocks are found, they usually turn out to have been altered by exposure to high temperature and pressure within the Earth. "On Mars there's no tectonics or mountain building or erosion," McKay reminds us. "We can see surfaces that are 3.8 billion years old—half the planet is that old. That doesn't mean that it's going to be easy to find fossils, of course. It may be hard in a different way."

If fossil microorganisms are found, that would still leave unresolved the question of whether they belonged to an independent Life, or were cousins to Schopf's terrestrial microbes. "We need the organic bodies," says McKay. To get them, he thinks, the most promising strategy would be to drill about a kilometer into the ancient permafrost surrounding the south polar cap. The idea would not be to get all the way down to liquid water but to find ancient frozen sediments that could harbor the remains of organisms dating from a more clement era. "They'd be dead," he says, "but we'd be able to see if they used the same twenty left-handed amino acids as we do, and DNA, and compare it to Woese's tree of life." With the DNA, of course, one would be halfway to a "Noachian Park." But given all the trouble that cloning dinosaurs led to, NASA may think twice about cloning Martians.

Currently, great efforts are made to sterilize Mars-bound spacecraft, to protect the martian surface from contamination with terrestrial organisms. But what if, after much searching, Mars is found to be lifeless? Some scientists and futurists have suggested that we should attempt to "terraform" Mars—to change its atmospheric conditions so that it could sustain terrestrial life and eventually become habitable by humans.[23] The initial steps in this transformation would have to involve physical and chemical methods, but according to Imre Friedmann, certain terrestrial microorganisms could be introduced quite early to speed the process of climatic change.[24]

Friedmann has already picked out two microorganisms as pioneers. One is *Chroococcidopsis*, a highly desiccation-resistant, cold-tolerant cyanobacterium that often lives as a hypolith under translucent pebbles. This organism could be grown on the martian surface under glass strips manufactured on site. Occasional wetting is all that would be required for the organism to carry out photosynthesis and liberate oxygen into the atmosphere. The other organism, an unnamed species of the genus *Matteia*, is also a desiccation-resistant cyanobacterium, but its special skill is to dissolve limestone, thus liberating carbon dioxide into the atmosphere. The carbon dioxide would supply *Chroococcidopsis*'s carbon needs and would also help warm the planet through the greenhouse effect. Together, Friedmann suggests, the two organisms could create a complete microbial carbon cycle and maintain a steady ratio of carbon dioxide and oxygen in the martian atmosphere.

Of course, terraforming other worlds raises ethical issues as well as technical ones. Some may see terraforming as the ultimate answer to terrestrial overpopulation. Others may see it as a destructive process, comparable to the environmental damage wrought by human activity on Earth.[25] If nothing else, terraforming illustrates the optimistic belief of some scientists that even current-day Mars is potentially habitable by organisms like some of those found on Earth today.

If Mars today lacks life (or hides it deep underground), what about other potential habitats in the solar system? The prospects are not very good. As we look inward toward the Sun, things get too hot. On Mercury, the innermost planet, the daytime surface temperature reaches 425°C; and Venus, with its thick greenhouse atmosphere of carbon dioxide, is even hotter: its mean planetwide temperature is about 460°C. As we look away from the Sun, on the other hand, things get too cold. Even Mars, as we have already discussed, is very borderline in that respect; and the next planet, Jupiter, has a mean temperature of minus 149°C. Our home planet seems to be perched in a "habitable zone" where McKay's crucial ingredient for life, liquid water, can exist on a planet's surface over a large fraction of its lifetime.

The acknowledged expert on habitable zones is Jim Kasting of Pennsylvania State University. His interest in the issue was sparked by the earlier work of physicist Michael Hart, who is now at Anne Arundel Community College in Maryland.[26] In the late 1970s, Hart did calculations suggesting that the habitable zone around the Sun was quite narrow, so that moving the Earth even a modest distance toward or away from the Sun would make liquid water an impossibility. And because the Sun has gradually been getting brighter, the region in which a planet could have possessed liquid water continuously over billions of years seems to have shrunk even further. This conclusion, along with other lines of argument that we'll discuss in later chapters, convinced Hart that life is extremely rare in the cosmos—so rare that we shouldn't waste our time looking for it.

However, Kasting (in collaboration with Dan Whitmire of the University of Southwestern Louisiana and Ray Reynolds of Ames) developed a model in which planets have feedback mechanisms that tend to stabilize liquid water.[27] The existence of these mechanisms greatly extends the habitable zone.

The key element of their model is carbon dioxide (CO_2). On Earth, CO_2 cycles between the atmosphere, where it exists as a free gas, and the Earth's crust and mantle, where it is chemically bound into carbonate rocks. The downward part of the cycle is dependent on liquid water, in which the CO_2 dissolves to begin the process of chemical combination; the upward part of the cycle is dependent on volcanism. Connecting these two processes, over time-scales of millions of years, is plate tectonics, which draws surface rocks down into the mantle, where temperatures rise high enough for the CO_2 to be driven off. Kasting pointed out that this cycle works like a thermostat: if the temperature drops, so that surface water freezes, CO_2 cannot pass from the atmosphere to the crust, but volcanism continues to pass CO_2 from the Earth's interior to the atmosphere. Therefore, the atmospheric CO_2 concentration goes up, and the resulting greenhouse effect heats the surface and melts the ice.

Kasting's model doesn't require carbon dioxide to be the only important greenhouse gas. As mentioned earlier, it may be necessary to invoke other gases, such as methane, to explain how the Earth and Mars possessed liquid water early in their history. But, Kasting believes, it is carbon dioxide whose levels are modulated to stabilize the climate in the face of changes in the Sun's brightness.

When Kasting slides his "model Earth" inward toward the Sun, it hangs on to its liquid water until its distance from the Sun is about 0.95 AU. (The real Earth is at 1.0 AU from the Sun, and Venus is at 0.72 AU.) At that distance, the atmosphere warms to the point that water vapor rises high into the stratosphere, where ultraviolet light breaks water molecules into hydrogen and oxygen, and the hydrogen escapes into space. After a few million years, this process exhausts the planet's entire inventory of water.

When Kasting slides the Earth outward, the carbon dioxide thermostat keeps water liquid until the Earth is about 1.37 AU from the Sun. (Mars is at 1.52 AU). Then it becomes so cold that carbon dioxide begins to condense out of the atmosphere, its contribution to greenhouse warming decreases, and temperatures tumble to the point that all surface water freezes. Kasting's original model could not explain how water ever existed on Mars, but the warming effect of clouds of solid carbon dioxide, as we mentioned earlier, may provide the answer, extending the habitable zone to include that planet.

Unlike Hart's model, Kasting's model is far more favorable to the notion that habitable planets are common in the universe, and in Chapter 5, we'll discuss the implications of Kasting's ideas for the quest for life beyond the solar system. But within our solar system, it would tend to rule out life anywhere but on Earth or Mars.

Not everyone thinks that we should abandon hope of finding life elsewhere in the solar system, however. The planets (and their satellites) have a habit of surprising us. In 1992, for example, radio astronomers Marty Slade, Bryan Butler, and Duane Muhleman, of Caltech and JPL, used radar to map the north polar region of the scorched planet Mercury—and found ice![28] The ice appears to be located on the floors of craters, where it is in perpetual shadow. Two years later, they found ice at the planet's south pole too. Then, in 1996, the Clementine spacecraft picked up radar signals suggestive of ice at the south pole of our own Moon. This finding was confirmed in late 1997 when the Lunar Prospector spacecraft found abundant hydrogen—probably in the form of water ice—at both poles.[29] These findings don't mean that life exists or ever existed on Mercury or the Moon, of course, but they show how global statements about conditions on planetary bodies may conceal a lot of local diversity.

One place that has been attracting a lot of attention is Europa, one of the four Galilean moons of Jupiter. Though it lies far outside the "habitable zone," there is strong evidence that Europa has possessed, and may still possess, an ice-covered ocean.

The evidence in support of this idea has come from images of the Europan surface transmitted by a series of space probes, culminating with the Galileo probe that is roaming among Jupiter's moons as we write.[30] Europa's surface is made largely of ice, and the ice is marked by numerous overlapping systems of linear tracks. In fact, Europa looks like a moonwide megalopolis in which freeway construction has gotten completely out of control. But there are no modern-day Lowells to interpret the scene in this fashion, and the cause of the tracks remains enigmatic. At any event, the pattern of tracks has been disrupted by the break-up of the icy crust into large blocks that have subsequently drifted or rotated from their original positions, as if they are floating on an underlying ocean. The gaps between the ice rafts have apparently filled with slushy water, which then froze solid. One can study the jigsaw images for hours, figuring out which blocks originally lay against which others, and how far they must have drifted or rotated.

Only an underlying liquid-water ocean, according to most re-

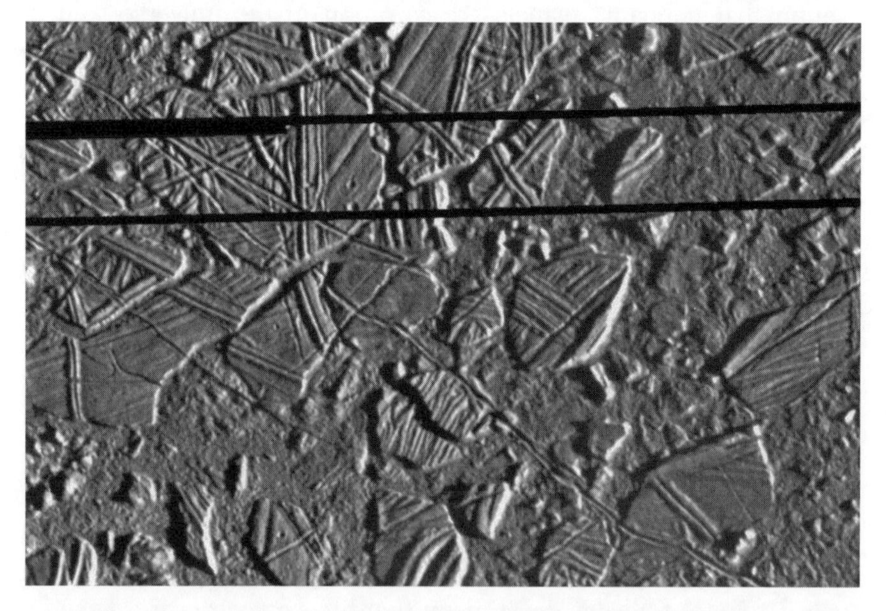

3.3 Ice-rafts on Europa, photographed by the Galileo spacecraft in February 1997. The area shown is about 42 kilometers wide. The two black bands represent missing data (JPL/NASA).

searchers, could have generated the current surface topography on Europa. What is not quite so certain is whether that ocean still exists today, or whether we are observing the petrified remains of a body of water that froze solid long ago. Because the surface generally looks very pristine, with very few impact craters, the majority opinion is that the ocean still exists.

Evidence in support of this conclusion was provided in 1998 by a UCLA/JPL/Caltech group that analyzed magnetometer readings obtained by the Galileo spacecraft. The researchers found that Europa, as well as another Jovian moon, Callisto, distorted Jupiter's magnetic field in a fashion best explained by the presence of liquid saltwater oceans on both moons.[31] Current estimates are that Europa's ocean is about 200 kilometers deep and is capped by 1 to 10 kilometers of ice—about the same thickness of ice that covers Lake Vostok in Antarctica (see Chapter 2).

Several methods could be used to further study these potential oceans. Ground-penetrating radar could detect liquid water beneath the ice. Repeated imaging could detect changes in the seascape as the ice rafts move. And laser altimetry could detect the up-and-down motion of the icy surface with the ebb and flow of tides in the ocean—tides that should be raised by Jupiter's enormous gravity. Future missions will use some or all of these techniques to provide an answer.

Europa's ocean remains liquid for the same reason that Lake Vostok remains liquid. The water is insulated by the thick layer of overlying ice, so it can only very slowly lose heat to the surface. This loss is compensated by geothermal heat coming from the interior. In Europa's case, the geothermal heat is generated mainly by tidal heating, as massive Jupiter flexes the body of its spinning satellite. It is quite possible that some of this heat passes into the ocean via hydrothermal systems similar to the deep-sea vents on Earth.

Of course, just possessing a water ocean doesn't mean that Europa is actually inhabited. Chris McKay, for one, is quite skeptical. "It's the next best bet to Mars," he says, "but it's not nearly as good a bet. There's no light under that ice. The only mechanism that could work is the chemosynthetic, hydrothermal one. But the good thing about Europa is that it's so far away. It's probably never been contaminated by life from Earth, or vice versa. If there's life there, it's probably a separate Life."

The long-term goal is to explore Europa's ocean directly (see Color Plate 4). Frank Carsey of JPL, who is one of the planners for this still-nebulous project, tells us that the present plan calls for the use of a "cryobot"—a radioactively heated probe that will melt its way down through the ice cap. The ice will refreeze above the probe, so it will be a one-way journey. Ice transmits radio signals poorly. Therefore, to maintain contact with the surface, the cryobot will release hockey-puck-sized microwave repeating stations into the ice every few hundred meters as it descends. These repeaters will be the cryobot's chain of communication with the surface—and thus with the orbiting mothership. When the cryobot reaches the ocean, it will release a robotic submersible, or "hydrobot," which will either be tethered to the cryobot or, more likely, will roam freely and return periodically with whatever information it can glean.

According to Carsey, with the help of the hydrobot, researchers will attempt to answer several questions. They will try to establish whether there are energy sources within Europa's ocean that could sustain life. They will look for out-of-equilibrium conditions that could result from living processes (Carsey mentions, by way of analogy, the simultaneous presence of oxygen and methane in the Earth's atmosphere—an unstable mixture that is sustained by the activity of the Earth's organisms.) They will also search for dissolved metabolites that might be produced by living organisms. And finally, they will try to find organisms themselves. This will be especially challenging, Carsey says. At one extreme, microbes are a problem because of their small size; while Loch-Ness-monster-sized creatures, at the other extreme, are notoriously shy.

One more object that deserves mention is mysterious Titan, the largest of the eighteen moons of Saturn. As far as its size (between that of Mercury and Mars) and atmosphere (denser than Earth's are concerned), Titan seems more like a planet than a moon. Its surface is hidden by a perpetual orange-tinted haze. The atmosphere is mostly nitrogen, argon, and methane. Solar radiation breaks methane into hydrogen and free radicals such as CH_2; the hydrogen escapes into space, and the free radicals promote the formation of larger hydrocarbons, which condense to form particulates—the cause of the haze. If this all sounds like Los Angeles on a hot summer's day, it is. But it's a lot cooler—the temperature at the surface is about minus 178°c.

The surface itself is thought to be partly solid: the Hubble Space Telescope, by imaging Titan in the infrared, was able to discern what seems to be an Australia-sized continent. The remainder, however, is probably liquid, and the favorite candidate for that liquid is ethane, probably with some other hydrocarbons mixed in. Underlying the ethane ocean, in some models, is a layer of solid acetylene. A dangerously flammable moon, one might think, but in the absence of oxygen these substances are, of course, harmless. And there is no oxygen because the most ready source of atmospheric oxygen—water—is completely frozen out of the atmosphere by the low temperature.

Carl Sagan, in the years before his death in 1996, became fascinated by Titan. In collaboration with his Cornell colleague Bishun Khare and others, Sagan attempted to replicate some of the atmospheric chemistry of Titan in the laboratory.[32] He subjected a simulated Titan atmosphere to irradiation by charged particles and obtained a dark brown organic gunk that he named "tholin," from the Greek for "mud." Over Titan's history, Sagan argued, huge amounts of tholin must have been deposited on its surface.

This finding alone makes Titan a potentially valuable natural laboratory for prebiotic processes. But Sagan also found that when exposed to water the tholin gave rise to amino acids, nucleotide bases, and many other building blocks of life as we know it. This was all the spark that Sagan's imagination required. Surely, impacts would have melted water ice on Titan's surface, he suggested, producing organic-rich lakes that lasted for millennia. Could life have actually arisen in these lakes, perhaps many times over, only to be brought to a standstill by the inevitable return of the deep freeze? Could the remains of that life still be there, waiting to be discovered?

Sagan's death prevented him from finding out, but the Cassini mission, now on its way to Saturn, may provide some of the answers.[33] The main Cassini spacecraft is expected to make at least thirty close flybys of Titan, starting in late 2004. During the flybys, the spacecraft will study atmospheric chemistry and the topography and makeup of Titan's surface. It will also release the Huygens probe, built by the European Space Agency, that will descend through Titan's atmosphere and land on who knows what. For as long as it may survive, rocking in a sea of ethane or sinking into a morass of tholin, the probe will send information about Titan's physical and chemical environment to Cassini, to be stored and later transmitted to Earth.

For all the current excitement about the possibility of life on Mars or Europa, there is a good chance that, of all the bodies orbiting our Sun, only Earth has ever been home to living things. For this reason, many scientists have pursued the quest for life, or for conditions favorable to life, among stars far beyond the Sun. It is too early to detect such life directly, unless it includes rational creatures that might send us radio messages. Yet there are several key questions about stars that are fundamental to the search for life elsewhere in the galaxy. Does the process of starbirth predispose to life by providing the molecules from which life can arise? And do planets regularly form around stars, or do we live in an unusual solar system with its nine planets, at least one of them habitable? If other planetary systems exist, are they like ours, or are they so different as to make life improbable? These questions form the topics of the following two chapters.

SUGGESTED READING

Christiansen, E.H., and W.K. Hamblin, *Exploring the Planets.* 2d ed. Englewood Cliffs, NJ: Prentice-Hall, 1995.

Kieffer, H.H. et al. eds., *Mars.* Space Science Series. Tucson: University of Arizona Press, 1992.

Sheehan, W. *The Planet Mars: A History of Observation and Discovery.* Tucson: University of Arizona Press, 1996.

4

The Death and Life of Stars

Organic Molecules and the Evolution of Solar Systems

From the dizzying catwalk that girdles the dome of the Hale telescope, we gaze out across the darkness of Palomar Mountain. Low on the southwestern horizon, a faint glow marks where San Diegans slumber—creationists, exobiologists, and those who care naught for the cosmos, all dreaming away the hours when California turns its back to the Sun. We too felt the pressure to sleep, half an hour ago, as we sat waiting idly in the telescope's warm control room. A computer sensed that the night air was too damp; fearing that dew might mar the mirror, we could not open the dome's portals, and our observations were delayed. So we bundled up against the winter chill and ventured outside for some pre-Galilean astronomy.

To city dwellers, the night sky at Palomar is always astonishing, and the view on this night is particularly fine. Along the ecliptic, the arc marking where the plane of our solar system meets the celestial vault, the four brightest planets march westward in a regular, stately procession: Venus, near to its setting, followed by ruddy Mars, Jupiter, and Saturn. Even without knowing of their peculiar motions, one can tell that these four "wandering stars" are not stars: they shine too steadily.

Sweeping across the ecliptic from the south to the northwest is the

Milky Way, our home galaxy. Its 300 billion stars merge into a broad river of light. Just a minute fraction of them, that happen to be close to us, form the myriad sparkling points that we call "the stars," yet even these few seem numberless. It's hard to doubt that, on a planet orbiting one of those stars, a living creature is gazing back at us.

To tell whether any of the stars really do have planets, and whether any of these planets are inhabited, seemed for centuries an impossible task. Even today, the world's greatest telescopes, such as the Hale telescope at Palomar, the twin Keck telescopes on Hawaii, and the Hubble Space Telescope, are incapable of spotting planets around other stars: not because the planets are too faint, necessarily, but because they cling too close to their bright parent stars. Yet a series of brilliant innovations has allowed astronomers to detect a handful of such planets indirectly and even to describe their basic properties, such as their size and orbital characteristics, as we'll describe in the next chapter.

Even more exciting than the detection of extrasolar planets, though, is our dawning understanding of the evolution of stars and their planetary systems, an evolution that may carry the seeds of life from the very beginning, even before a star and its planets are formed. Not many years ago, it was possible to believe that our Sun, with its nine planets—at least one of them inhabited—was a freak that might not be repeated within our galaxy. Now, though many details remain unresolved, we can speak of the birth of stars and their planets as one of the fundamental processes of the cosmos, and we can tentatively explain how the chemical building blocks of life take part in this process, making an extraordinary journey from deep space to end up in living creatures such as ourselves.

As we look up into the night sky, the winter constellation of Orion dominates the scene. The mighty hunter, followed by his dogs (Canis Major and Minor), chases Taurus the bull eternally westward. Orion straddles the celestial equator—the circle, oblique to the ecliptic, which lies directly over the equator of the Earth. Almost everyone recognizes this dramatic constellation and its brilliant individual stars with their exotic names: Betelgeuse, the variable red giant in Orion's shoulder; Bellatrix, in Orion's other shoulder; and the blue-white giant Rigel, the sixth brightest star in the sky, in Orion's foot. Most striking, though, are the three close-set stars that form Orion's belt and the fainter line of stars that dangle from it, forming his sword.

It wasn't till 1610, apparently, that anyone noticed something peculiar about the middle star in Orion's sword. A French lawyer,

Nicholas-Claude Fabri de Peiresc, saw that it was not a star at all but a fuzzy patch of light that extends, as we now know, for more than the Moon's width across the sky. It's a nebula. Many other sky gazers may have noticed the nebula earlier, perhaps even the Sumerians, though they were perverse enough to call Orion a sheep rather than a hunter. But Peiresc first wrote about it. Later Charles Messier, the great cataloger of nebulas, gave it a number: 42. So astronomers now call it M42 when they're in a formal mood, or the Orion nebula when they're not. But you don't need to be an astronomer to see it: it's easily spotted with the naked eye and a splendid sight through binoculars.

If you want to find extraterrestrial life, the Orion nebula is not the best place to look: it's mostly gas and dust, bombarded by intense radiation from four hot stars at its core. These stars, arrayed in a tight squarelike formation called the Trapezium, are tens of thousands of times brighter than our Sun and emit copious amounts of deadly ultraviolet radiation. They're also only about a million years old—far too young to possess habitable planets. But the Orion nebula is a stellar nursery, cradle to stars that may warm civilizations when life on Earth has long been extinguished. By studying what is going on in the Orion nebula today, and in places like it, astronomers hope to see stars and their nascent planetary systems, not simply as objects, but as examples of fundamental processes of Nature—processes that may, with a certain inevitability, lead to life (see Color Plate 5).

The Orion nebula is part of a much larger region of gas and dust called the Orion molecular cloud. This cloud is about 1600 light-years away from us (which still puts it in our corner of the galaxy), and it is several hundred light-years across, covering more than half of the entire constellation of Orion. It is one of several thousand molecular clouds in the Milky Way galaxy. These clouds form under the influence of spiral waves of compression and rarefaction that rotate around the galaxy's center as does the galaxy's constituent matter, but at half its speed. Thus, all the matter in the galaxy passes periodically through regions of increased density, which are visible as the galaxy's spiral arms. As the extremely tenuous gas and dust between the spiral arms approaches an arm, it speeds up; and then, as it crashes into the slower-moving material within the arm, it is compressed, forming molecular clouds. Gas in warm clumps within a giant molecular cloud may reach densities as high as a million molecules per cubic centimeter—but this is still 13 orders of magnitude less dense than ordinary room air.

The dense regions within molecular clouds are the sites of active chemical processes that could be thought of as the first steps toward life in the cosmos. Cosmic rays, energetic enough to penetrate the dust clouds, occasionally strip an electron from one of the atoms or molecules within the cloud. The resulting ions readily react to form more-complicated molecules, including many carbon-containing organic compounds, which can be detected by the characteristic patterns of microwave radiation that they emit. Protected from destructive ultraviolet radiation by the dust in the clouds, these molecules are deposited onto dust grains, forming icy mantles. Once frozen solid, they can no longer emit the microwave radiation that is detected by radio telescopes and therefore are not easily identified. By vibrating in place, however, they do absorb infrared light from background stars luminous enough to pierce through the dust. Each kind of molecule vibrates at characteristic frequencies, and absorbs infrared light that matches those frequencies. Thus, many constituents of the icy mantles are revealed by infrared absorption spectroscopy, although not as precisely as in emission spectroscopy of molecules in the gas phase.

Within the narrow confines of a dust-grain surface, further chemical reactions take place, including the formation of methanol and the most abundant molecule, molecular hydrogen (H_2). Here, too, are hydrocarbons in abundance, their production aided by ion bombardment and any ultraviolet radiation that can pierce the dust clouds. Among these compounds, polycyclic aromatic hydrocarbons (PAH's), similar to those found in the martian meteorite ALH 84001, coat the grains with a sludge not unlike that produced by decaying organisms on earth—a truly surprising find in deep space!

The core dust grains, containing perhaps a few million atoms each, are formed largely of silicon, carbon, and oxygen. These elements were not formed in the Big Bang, which produced little else than hydrogen and helium. Rather, they were formed by nuclear reactions inside an earlier generation of stars. When these stars reached the end of their existence, these elements were blasted out into deep space, where they gradually collected into grains.[1]

In the Orion nebula, the vaporizing power of radiation from massive stars eventually releases the organic molecules from the grains and returns them to space where they can emit microwave radiation and be detected by emission spectroscopy with radio telescopes. For this reason, the complete inventory of interstellar molecules is more easily assessed in "hot molecular cores" such as the Orion nebula.

Orion is a natural laboratory in which the organic products of dark clouds are newly exposed for study.

An amazing variety of organic compounds, including such household items as vinegar and alcohol, have been observed spectroscopically in the warmer regions of molecular clouds, and many more probably remain to be discovered. Some of these compounds, such as formaldehyde (H_2CO) and hydrogen cyanide (HCN), are the very molecules that have been credited with a crucial role in prebiotic synthesis.

Recently, a group of astronomers led by Jeremy Bailey of the Anglo-Australian Observatory made observations of the Orion molecular cloud that bear on the question of chirality—why terrestrial amino acids are all left-handed, for example. Bailey's group looked for circular polarization in the light coming from the cloud. Circular polarization means the direction of rotation—clockwise or counterclockwise—of a light wave's oscillating electric field. It has long been known from laboratory experiments that circularly polarized light selectively destroys organic molecules of one handedness, leaving an excess of molecules of the other handedness. Bailey's group found that infrared light scattered by dust grains in the Orion cloud can show very strong circular polarization. They could not make equivalent measurements in the ultraviolet, the region of the spectrum that is most effective at destroying organics, because the ultraviolet light is absorbed by the dense clouds of dust and never reaches Earth. According to the group's calculations, however, the ultraviolet light should also be circularly polarized to a degree sufficient to bias the chirality of organic compounds one way or the other. The direction of the bias depends in rather subtle ways on the geometry of the ultraviolet illumination and the precise wave bands in which the light is absorbed. Suffice it to say that different regions of a molecular cloud, and different classes of organic compounds within a single region, might end up with their chiralities biased in opposite directions.

At one level, the existence of interesting organic molecules in molecular clouds simply tells us how easily carbon forms complex molecules. So far, there is no hint of equivalent suites of molecules built up from atoms of silicon for example, the element that has often been proposed as the basis for an alternative life-chemistry (see Chapter 9), nor of any other fundamentally exotic classes of chemicals. Thus, one message of the molecular clouds may be that carbon is the natural choice for life everywhere.

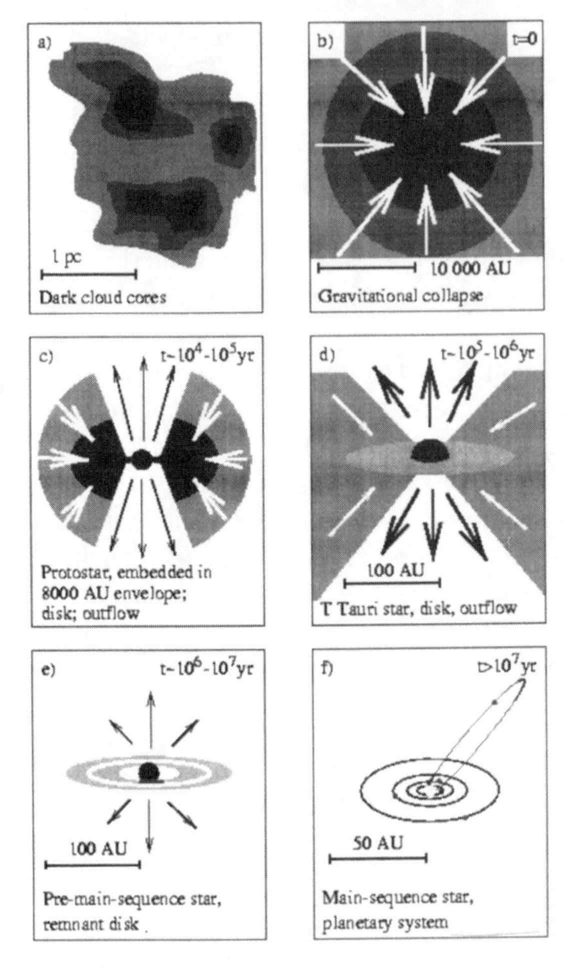

4.1 Schematic overview of the different stages of the formation of a low-mass star. (a) Local condensations in molecular clouds. (b) The onset of gravitational collapse and the formation of a protostar. (c) Continuing infall; the formation of a circumstellar disk and a bipolar outflow. (d) The T Tauri stage, characterized by accretion from the disk onto the star which is no longer obscured at optical wavelengths by the parental cloud core. (e) Slow contraction of the star to "adulthood," and coagulation of dust grains to form planetesimals in the disk. (f) An "adult" star surrounded by a planetary system. One parsec (pc) equals about 3.3 light years. One astronomical unit (AU) equals the mean distance between Earth and the Sun (about 93 million miles) (diagrams courtesy of Michiel Hogerheijde, adapted from Shu, Adams, and Lizano, Annual Review of Astronomy and Astrophysics, vol. 25, page 23, 1987).

We can also ask, however, whether there might be a more direct connection between the chemistry of life and the chemistry of molecular clouds. Could those very organic molecules actually travel from their birthplace in deep space and reach newborn planets, there to provide the ingredients for the first living systems? If this kind of migration occurs, we could make an even stronger case for the chemical relatedness of life throughout the galaxy. In addition, we might have an ultimate explanation for why terrestrial life uses left-handed amino acids, right-handed sugars, and so on. So it is of great interest to learn more about how solar systems form out of molecular clouds and what the fate of organic molecules in the clouds may be.

Gravity, of course, is the main player in the drama of starbirth. Within the Orion molecular cloud, tenuous though it is by earthly standards, slight irregularities in density are present, and over time gravity builds on these irregularities to the point that whole regions of the cloud begin to collapse inward on themselves. One such region of densification is a filamentous structure called the "Orion Ridge" (see Color Plate 6). Within the ridge is a chain of about a dozen denser spots, where self-shielding from the surrounding radiation has allowed the dust and gas to cool between 10° and 40°K.[2] These condensations of dust constitute true "protostars," the earliest phase in the formation of a star that we can observe. Most of a protostar's mass is still in the form of a cold dusty envelope falling onto a hidden stellar embryo. Protostars are too cold to radiate in the visible portion of the electromagnetic spectrum, or even at infrared wavelengths near the visible, but they do radiate at submillimeter wavelengths, the borderland between the radio and infrared regions. Protostars are probably no more than about 10,000 years old, a tiny instant in the life of a star.

No one has directly observed what is going on inside a protostar, but several astrophysicists have performed calculations or have made computer models that help to clarify a protostar's development. In the 1970s, Frank Shu of the University of California at Berkeley showed that the initial gravitational collapse of a molecular cloud core starts near the center (the site of the future star itself) and proceeds in an "inside-out" fashion.[3] Most of the matter that contributes to the star does so by first falling onto a flattened "accretion disk" that orbits the star rather like the rings that orbit Saturn.

Why does the material of a collapsing cloud core form both a star and an accretion disk, rather than collapsing directly into the star? The basic reason is that the original core of dust and gas from which the star forms is likely to be rotating, even if only at a very slow speed. Therefore it has some angular momentum. As the material collapses inward, the angular momentum must be conserved, and that causes the rate of spin to increase, as does a spinning ice-skater when she pulls in her arms and legs. If all the matter of the collapsing region were to form a single star, the star's rotation rate would have to be so great as to break it apart. Therefore, only a portion of the infalling matter contributes directly to the star; other material is left orbiting the star as a disk.[4]

According to theoretical studies conducted by several groups in the 1970s, the gas and dust in the disk should behave as a viscous fluid that requires energy in order to be stirred. The necessary energy is "stolen" from the material's angular momentum, allowing it to slow down, to spiral inward, and eventually to fall onto the star. In other words, the disk acts as an expanding conveyor belt that feeds material to the star from ever-greater distances. In the process, the "stolen" energy should heat up the disk, causing it to glow at infrared wavelengths. Another reason that the disk glows in the infrared is that it is warmed by radiation from the star.

These theoretical models received a powerful boost in the 1980s, with the launch of the Infrared Astronomical Satellite (IRAS). With this instrument, astronomers were able to detect infrared radiation coming from the centers of some molecular cloud cores. To distinguish such objects from younger protostars that emit only at longer wavelengths, astronomers dubbed them "young stellar objects." These are characterized by their near-infrared (near-to-visible) emission: the central star is emitting at visible wavelengths, but this light is absorbed by the surrounding dust and never reaches us. As more and more material is transferred from the cold, dusty envelope to the warmer star and disk, however, the system emits at shorter and shorter wavelengths, until optical light from the star can be directly detected—the star begins to "shine," in effect.

This relationship between temperature and emitting wavelength (or "color," in an expanded sense of the word) approximately follows Planck's law of blackbody radiation and is illustrated by the familiar appearance of an electric range as it warms up. At room temperature, the black coil of an electric range radiates only at infrared wavelengths and seems completely dark to the eye, but as it heats up, it shines dull red and then orange, corresponding to shorter wavelengths of emitted radiation. If raised to sufficiently high temperatures, the coil would continue to change colors (to yellow, for example).

We humans have evolved to see objects by virtue of the sunlight they reflect, and our eyes are only sensitive to the wavelengths at which sunlight is most intense. We cannot see most terrestrial objects in the glow of their own infrared radiation, but some cold-blooded animals such as pit vipers can see us in the infrared, thanks to specialized sense organs that we don't possess. The development of astronomical detectors such as those in IRAS, which extend human vision to that of pit vipers and far beyond, has led to a revolution in our understanding of the processes of star and planet formation, which take place at low temperatures.

Among these discoveries has been one that was not anticipated by any theory. As early as the protostar stage, broad rivers of molecular gas, as well as powerful, narrow jets of ionized gas, can be seen flowing away from the nascent star along the rotational axis of the protostellar disk (see Color Plate 7).[5] By this route, some fraction of the infalling material defies the grasping reach of gravity and escapes the system altogether. This surprising behavior is now explained as a further method of getting rid of angular momentum, probably aided by magnetic fields. By ejecting a small fraction of the mass at very high speeds, the young star can bring a larger mass of gas and dust inward to be used in building up the mass of the star.

In 1993, Robert O'Dell of Rice University pointed the Hubble Space Telescope toward the heart of the Orion nebula and obtained stunning views of the inside of a star factory (see Color Plate 8).[6] Scores of extremely young stars—just a few hundred thousand years old—are packed together at a density half a million times greater than the density of stars in our own neighborhood. At least half of them are surrounded by flattened disks or teardrop-shaped masses of orbiting dust and gas, most of them considerably wider than the diameter of our own solar system. These are made visible, not by the light of their own central stars, which are still relatively cool, but by the four bright Trapezium stars at the heart of the nebula, whose light ionizes the gases in them, making the disks glow. Other disks are seen in silhouette against the bright clouds behind. For the young stars nearest to the Trapezium, circumstellar material is being eroded and blown away by the intense radiation from the Trapezium stars—hence the teardrop or comet-tail shapes. The gas and dust may eventually be blown away entirely. But the evolving stars that lie at greater distances from the Trapezium have disks of regular shape that are likely to persist. These disks are thought to be the precursors of planetary systems.

Although the Orion nebula has yielded spectacular images of nascent stars, more has been learned about the details of star-forming systems in molecular clouds that are closer to us. The best-studied example is a cloud that covers parts of the constellations of Taurus and Auriga, in a region adjacent to Orion where the Milky Way crosses the ecliptic. The Taurus-Auriga cloud contains no bright high-mass stars like the Trapezium stars, and it therefore has no counterpart to the Orion nebula. To the naked eye, in fact, the cloud is only marginally recognizable as a dark void in the Milky Way, extending from just above the face of the bull (Taurus) to just below the squashed pentagon that represents the charioteer (Auriga). This apparently featureless abyss is really a region where the light coming from the stars of the Milky Way is blocked by dust in the Taurus-Auriga molecular-cloud complex. The complex is only about 450 light-years from us—a third of the distance to the Orion cloud. Here, telescopes that operate at longer-than-visible wavelengths can peer into the dark cloud and see the earliest phases of star formation up close and personal.

Unfortunately, the clarity with which details can be made out decreases with the longer wavelengths needed to penetrate the cold dust. The only recourse is to go to larger and larger telescopes. Ultimately, the size requirements go beyond what can be built or paid for. One way around this dilemma is to link an array of smaller telescopes in such a way as to afford the clarity of a single, much larger telescope. This technique is called interferometry, and will be described in more detail in Chapter 5. One such array, operating at millimeter wavelengths, is Caltech's Owens Valley Radio Observatory (OVRO) in the eastern Sierras near Bishop, California.

In 1986, using the OVRO array, Anneila Sargent of Caltech and Steve Beckwith (then at Cornell University) obtained the first image of molecular gas around an evolving star like our Sun.[7] They observed a flattened disk, twenty times the diameter of the solar system, around a young stellar object in the Taurus molecular cloud known as "HL Tauri." or "HL Tau" (with star names, the constellation is often abbreviated to its first three letters). Its light emerges indirectly, reflecting off the walls of a cavity in the surrounding dust that was blasted open by outflowing gas. This reflection was originally mistaken for the star itself until the Hubble Space Telescope revealed it for what it was. Sargent and Beckwith originally believed they'd found evidence that all the molecular gas they detected was rotating around the star as a very large protoplanetary disk. More recent observations, however, suggest that much of the gas in the outer disk is still falling inward onto a somewhat smaller rotating disk. This implies that infalling material is mixing into the outer edge of the disk rather than impacting onto the surface of the disk nearer to the star.

Such observations are crucial to the question of whether the organic compounds found on the grains in the original molecular cloud could ever make it into an evolving planetary system. In Stanley Miller's view, the compounds would be destroyed by the "accretion shock" as the grains fell onto the protostellar disk. But Jonathan Lunine, of the University of Arizona's Lunar and Planetary Laboratory, has developed a model suggesting that the fate of the grains, and the organic compounds they carry, depends on the distance from the star at which the infalling grains meet the disk.[8] At 1 astronomical unit (corresponding to the Earth's distance from the Sun), particles do indeed smash into the disk at speeds sufficient to vaporize even tough silicate grains and to break down all organic molecules. Beyond 1 astronomical unit, however, impact velocities are not high enough to completely vaporize silicate grains, and beyond 5 astronomical units (the orbital distance of Jupiter) even water-ice grains are not fully vaporized.

In addition to impact velocity, the amount of heating and chemical processing of the organic mantles of the grains turns out to be very sensitive to the grain size, initial grain velocity, and details of the accretion shock itself. But for a wide range of assumptions, grains that enter the accretion disk at distances of 30 astronomical units or more (the orbital distance of Neptune) do not suffer appreciable vaporization of their organic mantles, much less any breakdown of the constituent molecules. Comets and icy bodies, like Pluto or the "Kuiper-belt objects" (see below), may have incorporated a large reservoir of primordial interstellar organics, like those detected in the Orion cloud. Indeed, the European Space Agency's Giotto spacecraft, that flew by Halley's comet in 1986, detected a surprisingly rich inventory of organics.[9] Thus, it would seem that organic molecules can survive the trip into an evolving planetary system, but only at initial entry points far from the star. So it becomes important to know whether disks typically grow to large diameters, such as 30 astronomical units, or not. To answer this question, it's necessary to study mature disks that are no longer surrounded by an envelope of free-falling material, but that consist mostly of material that is prevented from falling inward by the material's orbital motion within the disk.

The young stars of this kind in the Taurus-Auriga cloud, and similar ones in other clouds, are named "T Tauri" stars, after the particular star T Tauri from which unusual emission and surrounding nebulosity was identified as early as 1945. These stars have about the same mass and surface temperature as the Sun, or less, but they are brighter. This is because they have not yet condensed down to their adult or "main-sequence" size and therefore have a larger surface area available to radiate away photons. T Tauri stars still derive their energy largely from their gravitational collapse. Although the center of a T Tauri star may be as hot as 1,000,000°K, that is still not hot enough to begin hydrogen fusion, the primary step in the sequence of nuclear reactions that power main-sequence stars. Besides the visible-light radiation, T Tauri stars also radiate profusely at x-ray wavelengths. Both the visible and the x-ray radiation varies over time, as if giant "star spots" are forming and dissolving on the stars' surfaces.

Although not surrounded by as much dust as young stellar objects, many T Tauri stars radiate more powerfully at infrared wavelengths than one would expect on the basis of their surface temperatures. This excess infrared radiation was first detected in the 1960s and was correctly interpreted as evidence for the presence of a dust disk surrounding the star. Visible light from the central star warms the dust, and the dust then reradiates the energy at infrared wavelengths. If the dust were configured in a complete shell around the star, the starlight itself would be obscured, as it is for young stellar objects; therefore, the dust had to be in some kind of flattened, disklike shape.

One of us, David Koerner, has been particularly involved in studying the evolution of circumstellar disks. Koerner wanted to know whether the material around T Tauri stars is indeed in a stable orbit or is simply falling toward the star. Using the OVRO array, Koerner, together with Sargent and Beckwith, was able to image the radio emissions from cold carbon-monoxide gas in the disk of a T Tauri star called "GM Aurigae." He found that the radio emissions were Doppler shifted to longer wavelengths on one side of the disk (meaning that this side is moving away from Earth) and to shorter wavelengths on the other side of the disk (meaning that this side is moving toward Earth). In addition, computer simulations of images taken at the intervening velocities showed that the gas moves around the star with a speed that decreases with distance in exactly the manner predicted for orbiting bodies by Kepler's Laws. Thus, the whole disk is indeed rotating in a stable orbit around the star and therefore could easily provide the parent material for a set of planets.

The shape and orientation of GM Aurigae's disk was recently brought more sharply into focus by the Hubble Space Telescope. Using a near-infrared camera equipped with a mechanism called a coronagraph that blocks out much of the light from the central star, Koerner and others confirmed the disklike shape and the orientation assumed in the interpretation of OVRO images. It turns out that the gaseous disk is actually several hundred astronomical units in radius, several times larger than the orbital radius of Pluto. Koerner and Sargent have since imaged many more T Tauri stars, with the same result: their disks typically have radii several times the radius of Pluto's orbit. Thus, much of the organic material that enters into the disks is likely to do so at great distances from the star, where the "accretion shock" is so slight as not to threaten the material's chemical integrity.

More recently, Koerner and Sargent have detected the telltale signature of a rotating disk in the carbon monoxide signal from "LkCa 15," also in the Taurus-Auriga cloud. LkCa 15 is a borderline example of a so-called "weak-line" T Tauri star, which is commonly believed to represent a later stage of stellar evolution in which the dust of the circumstellar disk is beginning to thin out. Perhaps more exciting for our quest, HCN was recently imaged in the disk around LkCa 15 by Charlie Qi and Geoff Blake of Caltech. Here is a molecular component of great importance to Urey-Miller prebiotic synthesis, actually detected within a protoplanetary disk! A broad census of molecules in disks like those in the Orion nebula, however, awaits instruments of higher sensitivity.

Just as molecules move from the gas phase to grain surfaces in cold dark molecular clouds, they probably also freeze out onto dust grains in the dense environment of circumstellar disks. It thus becomes increasingly difficult to detect the weak microwave signature of emission from the few remaining molecules in the gas phase. Consistent with this picture, infrared absorption spectroscopy has revealed hydrocarbons in grains within young stellar objects. In another case, infrared imaging with the Keck telescope revealed that a disk around the young stellar object, "WL 16," is glowing throughout with PAH radiation. All in all, it would seem that we are on the threshold of discovering that the milieu in which planets form is rich in organics.

When T Tauri stars reach an age of about 10 million years, they are approaching "adult" or "main-sequence" status, when their cores will shine solely by light generated by the fusion of hydrogen. T Tauri stars of this age radiate little or no excess infrared light, suggesting that the smallest dust grains in their disks have largely disappeared. What then is the fate of the grains and their rich organic freight? One possibility is that they are simply blown away into space. Another more interesting possibility, however, is that they aggregate into larger bodies, some of which may move inward from the disk's peripheral zone, where they originally accreted to the inner zone where rocky Earth-like planets might form.

George Wetherill, of the Carnegie Institution in Washington, DC, has approached this question by modeling solar systems on a computer. He starts by playing a sophisticated version of the once-popular video arcade game "Asteroids," with a large population of orbiting "planetesimals"—these are the roughly 100-km-size bodies that are believed to be the building blocks of planets in the late stages of disk evolution. Rather than blast them with photon torpedoes, Wetherill simply lets them collide with each other under the influence of gravity. In this way, he makes terrestrial planets with ease and, by adding a Jupiter in the right place, he prevents any planet from forming in the vicinity of the asteroid belt, just as seems to have happened in our own solar system.

In the outer part of the model solar system, however, planet formation proceeds very reluctantly. Beyond Neptune's orbit, the average time between collisions becomes as long as the present age of the solar system. So Wetherill is left with a population of cold icy planetesimals that still occasionally collide but that never completely accumulate into a single planet.

These objects do in fact still exist in the outer reaches of our own solar system. They're called "Kuiper-belt objects," after Gerard Kuiper, who first postulated a family of comets orbiting in this region. Examples of such icy bodies have recently been detected by a number of researchers, especially by Jane Luu of Harvard and Dave Jewitt of the University of Hawaii.[10] The objects bear a striking resemblance to Pluto and its moon Charon, as well as to Neptune's captured moon Triton. Of importance to us is the fact that the objects form in the outer region of the developing solar system where interstellar organic compounds can survive the accretion shock.

When the solar system was young, according to Wetherill, there should have been many more rocky and icy planetesimals than the current inventory of asteroids and Kuiper-belt objects. Some of the missing objects, his model indicates, were slung out of the solar system after near encounters with planets or with each other. Judging by the cratering record on the oldest known planetary surfaces, however, an astonishing number were not ejected from the solar system but wended their way to a fiery collision with one or another planet or moon, including Earth. The recent collision of comet "Shoemaker-Levy 9" with Jupiter serves to remind us that such events continue even today.[11]

Could infalling comets and icy planetesimals, packed full of organic molecules, have contributed substantially to the origin of life on earth? We already visited this subject in Chapter 1. According to Chris Chyba and the late Carl Sagan, the inflow of organics from space could have been as high as the local (Miller-style) production rate in a weakly reducing atmosphere. Chyba also points out that the inflow could have been even greater if the early atmosphere was substantially thicker, as some believe it to have been. A 10-bar atmosphere (ten times the pressure of Earth's current atmosphere), by exerting greater drag on infalling objects and removing heat from them more efficiently, would have preserved the organic freight of small comets, up to about 100 meters in diameter, thus allowing entry of a particularly rich source of prebiotic chemicals.

Currently, a number of astronomers are studying the final stages of stellar maturation, hoping to find clues to the eventual fate of circumstellar disks and the possible formation of planets. In 1998, for example, Koerner (along with colleagues Michael Ressler, Michael Werner, and Dana Backman) equipped the Keck telescope with a new long-wavelength camera (built by Ressler) and pointed it at a 10-million-year-old star named "HR 4796," which lies about 220 light-years from Earth in the southern constellation of Centaurus.[12] When the telescope was first aimed at the star, no image appeared on the computer screen, to the astronomers' dismay. Thinking that there might be a problem with the pointing of the telescope, they searched around in the vicinity of the star, but still could not find it. It then occurred to them that, if most of the emission were spread out in a disk rather than concentrated near the star, the emission might appear weaker than expected. Acting on this hunch, Koerner and his colleagues increased the exposure time, and finally a ghostly image appeared on the screen (see Color Plate 9). A rush of excitement filled the control room as it became apparent that the team had imaged a circumstellar disk around the star. By complete coincidence, two other teams made similar observations of HR 4796's disk within a day or two of the Keck group: a team led by Bradford Smith of Hawaii's Institute for Astronomy photographed the disk with the Hubble Space Telescope, and a group from Harvard University and the University of Florida (led by Ray Jayawardhana) photographed it from an observatory in Chile.[13]

The disk around HR 4796 measures about 150 astronomical units across, which is about twice as wide as our own solar system. Most interestingly, however, the disk seems to have a central hole about the size of our solar system. The most plausible explanation for the hole is that the material in this region has already collected together into larger objects, such as kilometer-sized planetesimals, or even into full-sized planets. If planetary bodies of some kind were not present, radiation from the star would long ago have caused the ring to spread out, fill in the hole, and ultimately disappear, since its gas and dust would fall into the star or be blown into deep space. Some unseen influence must maintain the concentration of dust in the outer disk and keep the inner region relatively clear.

Indirect evidence that planetary systems do indeed surround main-sequence stars was provided by the study of dust emission from stars like Vega, the brilliant chief star of the constellation Lyra. With a mass several times that of the Sun, Vega burns its nuclear fuel at a faster rate, so it shines brighter and has a higher temperature. Its circumstellar dust is thus more strongly heated, and it emits more infrared radiation for a given mass of dust, compared with the dust around a cooler star. Using the IRAS satellite and other techniques, astronomers have detected excess infrared radiation coming from many Vega-type stars with ages of a few hundred million years. The disk around one of these, Beta Pictoris, was imaged in visible light in 1984 by Brad Smith (then at the University of Arizona) and Richard Terrile (of JPL), using a telescope in Chile equipped with a coronagraph. Two circumstances helped them achieve this feat: the disk is very large—more than 400 astronomical units across—and it is seen almost exactly edge-on.

At first, Beta Pic's disk was thought to be a protoplanetary disk—the structure from which planets might form in the future. But further observations have made it seem more likely that it consists of the leftovers from the already completed process of planet formation. For one thing, the total amount of dust remaining in the disk is very small—less than half the mass of Earth. Also, a central hole, about the size of the solar system, has been inferred on the basis of the spectral distribution of the infrared emission. As with HR 4796, its presence could be explained by clearing due to planet formation. Finally, the disk is distorted: in 1996 Chris Burrows (of the Space Telescope Science Institute in Baltimore), with the Hubble Space Telescope, obtained a photograph of Beta Pic showing that the part of the disk nearest to the central hole is tilted out of the plane of the rest of the disk, and a more recent HST image by Sally Heap showed the disk's warp even more clearly (see Color Plate 10).[14] According to Burrows, the tilt is probably caused by the presence of a roughly Jupiter-sized planet orbiting in the central clear zone, near to the inner margin of the dust disk. If this planet's orbit were tilted by a few degrees out of the plane of the disk, it would cause the observed tilting of the disk. Given that other mechanisms might have produced the warping, however, the hypothesized planet remains just that—hypothetical. Direct detection of extrasolar planets, where life might not only originate but evolve and thrive, requires more specialized techniques, as we'll see in the next chapter.

The disk around Beta Pictoris is now usually described as a "debris disk," implying that it consists, not of the original gas and dust that collapsed to form the star, but of material thrown off during collisions between asteroids, planetesimals, or other objects. While the predominant infrared radiation comes from micron-sized particles, the disk may well contain many larger rocky or icy objects. In fact, it may resemble the Kuiper belt that encircles our own solar system.

So far, the observations on young stars support the notion that planets form by gradual accumulation of smaller objects within a circumstellar disk. The disk model also seems to help explain the particular kinds of planets that we see in the solar system and their relative positions. There is likely to be a gradient of temperature within a disk, with the inner regions being hotter than the outer. Thus, within the inner regions, only rock and metals can condense and form the objects that later aggregated to form planets. This offers an explanation of why Mercury, Venus, Earth, and Mars are all relatively small, rocky planets. In the frigid outer regions, on the other hand, ices can also condense, adding greatly to the mass of the planets forming there. Those planets are also able to grow larger simply by virtue of their longer orbits, which contain more material available to be swept up. And once past a certain mass, the outer planets can attract and hold onto a dense atmosphere of hydrogen and helium, derived from the disk. This is why the outer planets—Jupiter and Saturn especially—became the solar system's "gas giants."

Is this, then, the pattern we should expect to see repeated over and over again across the galaxy? Are myriad suns surrounded by clones of our own solar system? Gazing at the night sky from the dome of the Hale telescope at Palomar, we can only ponder the question, as so many have pondered before us. But technology is finally closing in on the answer. Part of that technology lies housed below us in the darkness: the Palomar Testbed Interferometer, the prototype for an astronomy of the future, hugs the mountaintop like a colossal three-legged spider. Within its hublike central building, technicians are putting starlight through an optical maze, hoping to squeeze celestial secrets out of it.

SUGGESTED READING

Koerner, D.W. "Analogs of the Early Solar System." *Origins of Life and the Evolution of the Biosphere* 27 (1997): 157–84.

Mamajek, E. "T Tauri Stars: 'Young Suns'" [available at http://www.ph.adfa.edu.au/e-mamajek/ttau/index.html].

5

The Planet Finders

Searching for Life Beyond the Sun

Detecting extrasolar planets is difficult, and nowhere is that difficulty better illustrated than by the "holes" in the disks of stars like Beta Pictoris and HR 4796. Matter finely spread out, as in a disk, can radiate and reflect photons on a grand scale; but bundle that same matter into planets, as may have happened in those holes, and most of the matter is hidden from view, unable to communicate its presence by optical means. The light that planets do emit is likely to be swamped by the far more intense radiation from their parent stars. How, then, can we detect planets? The efforts devoted to doing so, which have been crowned with success only within the last 10 years, say a lot about our yearning to find Earth's peers in the galaxy.

One simple way that planets might make their presence known is by partially eclipsing their parent star. Some stars do periodically dim for a brief interval: a well-known example is Algol ("the ghoul" in Arabic), the second-brightest star in the constellation Perseus. Until well into the twentieth century, astronomers thought that Algol's dimming was caused by a large planet, whose orbit happened to intersect the line of sight between Algol and Earth. It's now known that Algol is indeed periodically eclipsed, but by a fainter stellar companion, not by a planet. Still, this approach to detecting planets may yield genuine results in the future, as we'll discuss later.

Another possible avenue to detecting planets depends on planets' gravitational interaction with their parent stars. Although we normally speak of planets as if they orbit their star, that's not quite right. In reality, the star and all its planets orbit the center of mass of the entire system. As the mass of the planets is unevenly distributed around the star, the center of mass will not coincide with the center of the star, and so the star will wobble like an unevenly loaded clothes dryer. Detecting that wobble would be equivalent to detecting the planets that cause it.

Unfortunately, the motion of the star induced by the presence of planets is likely to be quite small. The Sun, for example, is caused to wobble primarily by giant Jupiter, which possesses 70 percent of the mass of the entire nine planets. Even Jupiter, however, has only about one-thousandth the mass of the Sun, so the center of mass of the Sun-Jupiter system is a thousand times closer to the center of the Sun than it is to the center of Jupiter. This point is barely outside the surface of the Sun itself. As seen from even the nearest star, the total side-to-side wobble of the Sun, caused by the presence of Jupiter, would measure only about 4 milliarcseconds (an arcsecond is 1/360 degree, and a milliarcsecond is one-thousandth of that). That's equivalent to the separation of the two headlights of a car, as seen from a distance of 20,000 kilometers. And the other, less massive planets induce even smaller wobbles.

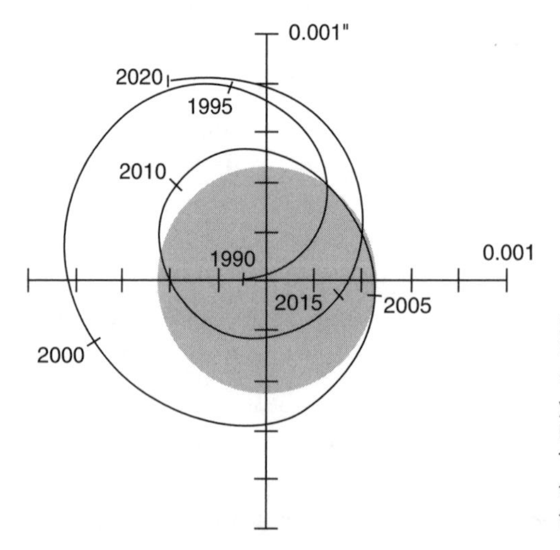

5.1 Apparent motion of the Sun due to its planets over a 30-year period, as seen from a distance of about 32 light years. Most of the wobble is caused by Jupiter. The scale is in arc-seconds. The central shaded area represents the diameter of the Sun as seen from that distance (JPL/NASA).

Conceptually, the easiest way to detect a star's wobble is to measure changes in the star's position in the sky over a period of time. This so-called astrometric method was already used successfully in the mid–nineteenth century to detect the presence of a companion to brilliant Sirius, the Dog Star. But Sirius's companion is another star, not a planet. In spite of extraordinary technical refinements over the years, there has not yet been an unambiguous detection of an extrasolar planet with the astrometric method. In 1996, the chief devotee of astrometry, George Gatewood of the University of Pittsburgh's Allegheny Observatory, did produce evidence suggestive of one or two planets around "Lalande 21185," a red dwarf that is the fourth-nearest star to the Sun.[1] Gatewood's claim has been greeted with fairly widespread skepticism, however.

A less obvious but (as it turns out) more fruitful approach to detecting a star's wobble involves measuring its motion along the line of sight between the star and the Earth—its so-called radial velocity. If a planetary system is viewed edge-on, the central star will move alternately toward and away from Earth as it orbits the system's center of mass. In fact, there will always be some component of radial motion, except for the unlikely case that the system is viewed exactly pole-on.

Comparing the radial-velocity approach with the astrometric approach, one can see that they are biased toward detecting different kinds of planets. Both methods work best on large planets. But the astrometric technique most readily detects a planet that is orbiting at a great distance from its parent star, because such a planet produces the greatest side-to-side excursion of the star. The radial-velocity approach, on the other hand, works best for a planet that is close to its star, because such a planet travels at a greater speed in its orbit, leading to a greater velocity of the star toward and away from the Earth.

In 1991, Alex Wolszczan (pronounced VOLL-shun) detected the first extrasolar planets.[2] The Polish-born astronomer was working at the giant Arecibo radio telescope in Puerto Rico, where he was studying pulsars—sources of regular pulses of radio energy. Pulsars are rapidly rotating neutron stars, the highly condensed remnants of giant stars that have ended their existence by exploding as supernovae. A pulsar emits a beam of radio energy that sweeps across the sky as the star rotates. If the beam happens to intersect Earth, we can detect the beam as a series of pulses.

The particular pulsar that Wolszczan discovered, named "PSR B1257+12," lies 1300 light-years away in the constellation Virgo. It rotates once every 6.2 milliseconds, or 642 times per second. (Even at this phenomenal speed, it is prevented from flying apart by its enormous gravity.) Wolszczan noticed that the timing of the pulses was not quite regular. They would speed up very slightly for a few days, then slow down again. He wondered whether movement of the pulsar toward and away from the Earth, induced by a planet or planets, was causing the oscillations in pulse frequency. The changes in frequency would then be Doppler shifts, indicative of changes in radial velocity.

With the assistance of Dale Frail of the National Radio Astronomy Observatory at Socorro, New Mexico, Wolszczan was able to prove his hypothesis correct. The frequency oscillations were accounted for by giving PSR B1257+12 two planets, both quite close to their star. The inner planet (0.36 AU from the pulsar, about the same as Mercury's distance from the Sun) has an orbital period of 66.5 days. The outer planet (0.47 AU from the pulsar) has an orbital period of 98.2 days. The planet's masses cannot be exactly calculated, because it's not possible to determine the angle at which we're viewing the planetary system. If we're seeing it exactly edge-on, the two planets have masses of 3.4 and 2.8 Earth masses, respectively. If the plane of the system is tilted away from edge-on, the masses will be greater, but probably not radically so.

A couple of years after his original discovery, Wolszczan announced that PSR B1257+12 had yet a third planet, this one a Moon-sized object orbiting even closer to the pulsar than were the two planets previously detected.[3] According to Wolszczan, the object orbited the pulsar once every 25.3 days. Recently, however, a German astronomer, Klaus Scherer of the Max Planck Institut für Aeronomie in Katlenburg-Lindau, has reported detecting the same 25.3 day Doppler oscillation in radio signals coming from Pioneer 10—a spacecraft that, having completed its survey of the outer planets, is now heading out into deep space. Scherer believes that a single source must account for the oscillation of the signals from both the pulsar and the spacecraft.[4] The source must be local to our solar system and is probably related to the rotation of the Sun. As if to compensate for the apparent loss of his innermost planet, however, Wolszczan has come up with preliminary evidence for yet another planet around PSR B1256+12, this time a Saturn-sized object orbiting far beyond the other two.[5] (Two researchers at MIT, Kriten Joshi and F.A. Rasio, have done independent calculations that support the existence of this planet.[6])

Wolszczan's discovery was an exceptional achievement and made astronomical history. But it wasn't quite what was wanted for those who seek planets because of their potential for life, because a planet circling a pulsar is not a good place to live. If the pulsar beam hits the planet, the harsh radiation (gamma rays and charged particles) will surely prevent any life from developing. If the beam doesn't hit the planet, it will be cold and dark. And because the search for life is in most people's minds, PSR B1257+12 has tended to be ignored. Oftentimes, the discovery is alluded to merely by exclusion, as in statements like "The first detection of a planet orbiting a normal star was ..."

The first detection of a planet orbiting a normal star was achieved by Didier Queloz and Michel Mayor of the University of Geneva, Switzerland.[7] Queloz was a graduate student, and Mayor his advisor, when they made the discovery in 1995. Queloz is now a postdoctoral fellow at JPL working on the Palomar Testbed Interferometer, so we have little trouble tracking him down.

"I wasn't supposed to find planets," he tells us in a charming French-Swiss accent, over a tiny cup of espresso coffee. "Everyone expected that any planets we could detect—the big ones—would be far from their stars, like Jupiter, so they would take 10 years or more to complete one orbit. But 10 years is not the kind of time scale you can think of when you are a graduate student. So I was just supposed to start the program, write the software, and debug the instrument."

The instrument in question was a spectrometer. Queloz and Mayor wanted to detect oscillations in the radial velocity of a star (induced by a planet or planets) by measuring the Doppler shifts in the wavelength of light coming from it: a red shift as the star moved away from the Earth and a blue shift as the star moved toward the Earth. The problem was that the shifts were likely to be minute. Jupiter, for example, causes the Sun to move to and fro at a speed of about 12 meters per second—only slightly faster than a sprinter can run. Compare this with the speed of light—300,000,000 meters per second. If a star is orbited by a Jupiter-like planet, the wavelength of the light reaching Earth will be alternately shortened and lengthened by the fraction 12/300,000,000, or 0.000004 percent. For an Earth-like planet, you'd have to add another couple of zeroes to that percentage.

The light coming from a star is spread out over a wide range of wavelengths, but this broad spectrum is interrupted by thousands of absorption bands, the signatures of the many elements and compounds present in the star's outer layers. It was the collective shift of all these bands toward longer or shorter wavelengths that Queloz and Mayor set out to measure. They did this by comparing the star's absorption bands with the absorption bands in the light coming from a lamp, which obviously were not Doppler shifted.

Several other astronomers had been using the radial velocity approach for several years before Queloz and Mayor commenced their observations. These others included two Canadians, Bruce Campbell and Gordon Walker of the Dominion Astronomical Observatory in Victoria, as well as two astronomers at San Francisco State University, Geoff Marcy and Paul Butler (Butler is now at the Anglo-Australian Observatory in Sydney). Aside from a couple of false alarms, none of these scientists had succeeded in detecting any planets. What Queloz and Mayor had going for them, however, was a sophisticated piece of analytical software that very rapidly calculated the Doppler shift of the source while the star was still under observation. Marcy and Butler's spectra, on the other hand, had to be processed off-line, and in fact they had accumulated several years' worth of unanalyzed spectra when the Swiss researchers began their observations.

Queloz and Mayor chose about 125 stars to study. They had masses similar to that of the Sun (so-called G stars) or slightly lower (K stars). Planets of Sun-like stars are often considered the best candidates for harboring life, so we ask Queloz whether the quest for life was a factor behind their choice of targets. He denies it. "They're just the easiest stars to study," he says. "The lower-mass stars are less bright, and the higher-mass stars don't have enough absorption bands to study."

They set up their equipment at the Observatoire de Haute-Provence, just outside the medieval town of Forcalquier in the French Alps. Forcalquier's Romanesque Church of Notre Dame dates back to the era when Scholastic philosophers attempted to solve the problem of "other worlds" by the power of thought alone. Queloz and Mayor resolved the problem empirically in July 1995, when they detected a large planet circling a star called "51 Pegasi" (usually abbreviated to "51 Peg"). It is the fifty-first in a list of stars in the constellation Pegasus.

Queloz and Mayor had actually first noticed the variation in the radial velocity of 51 Peg six months earlier. But the data were so unexpected that they hesitated to believe them. For 51 Peg was moving to and fro with a periodicity of only 4 days, and yet the amplitude of the velocity change demanded a planet at least half the mass of Jupiter. If the data were real, it meant that a Jupiter-like planet was orbiting its star at a distance of only 0.05 AU—one-eighth of the distance at which Mercury orbits the Sun, and in fact only about 4 stellar diameters from 51 Peg's surface. This was in radical contradiction to the standard model, which asserted that giant planets should lie far from their stars, as they do in our own solar system.

A few weeks after they had begun their observations, the advancing seasons turned 51 Peg into a daytime star, and Queloz and Mayor had to cool their heels for 4 months. But they used the time to further analyze their data and to predict the star's velocity on the night in early July when they would next be able to observe it. For even though the candidate planet would have completed nearly thirty orbits in the meantime, its orbital period was known precisely enough to predict its exact position in its orbit several months in the future. When that night arrived, 51 Peg had exactly the predicted velocity. After ruling out several possible artefactual causes for their results, Queloz and Mayor had their planet in the bag. In early October, Mayor announced the discovery at a conference. Marcy and Butler, though initially skeptical, were able to confirm the discovery by observing 51 Peg's radial-velocity oscillations themselves at the Lick Observatory in Northern California.

Since being scooped by Queloz and Mayor in 1995, Marcy and Butler have been making most of the running. Not only have they confirmed the planet around 51 Peg, they also have reported thirteen planets of their own. Four of these planets, which orbit stars named "55 Cancri A," "Tau Boötes A," "Upsilon Andromedae," and "HD 187123," are similar to 51 Peg's planet—that is, they are roughly Jupiter-sized bodies orbiting very close to their star. Yet another planet of the same type has been discovered by a team from the Harvard-Smithsonian Center for Astrophysics and two other institutions, led by Robert Noyes: this planet circles a star named "Rho Coronae Borealis."

The fact that many of the recently detected planets have been "close-in Jupiters" reflects in part the intrinsic bias of the radial-velocity technique, which is most sensitive to stellar motions induced by large, rapidly orbiting planets. But the existence of any such planets presents a significant challenge to the standard model of planet formation. One way to preserve the standard model is to hypothesize that the close-in Jupiters formed much farther from their stars, but then migrated inward. In fact, Doug Lin of the University of California, Santa Cruz, proposed exactly such an inward migration in the 1980s, long before any extrasolar planets had been detected.

We have the opportunity to ask Marcy about this when we visit him at his office at the University of California at Berkeley, where he has an adjunct appointment. The walls of the office are festooned with what look like seismographic tracings but are really the output of his spectrograph. "Planets move inward because the whole protoplanetary disk is going down the stellar drain," he says. "The disk is in fact making the star—that's how the star comes into being. The disk is slowly, over the course of a million years, swirling inward, dragging the planets with it. It's like when you take a bath and drain the water—the viscosity of the swirling water robs the water of its speed and causes it to spiral down the hole."

This, then, leaves two questions: Why has our Jupiter not done the same, and why have the close-in Jupiters not fallen right into their stars? As to the first question, Lin believes that several gas giants may have formed in the distant reaches of our solar system, migrated inward, and fallen into the Sun. Jupiter was the last, and it got left high and dry when the material of the disk was finally all swept up.

As to why the close-in Jupiters haven't fallen into their stars, we ask Marcy about the theory he espoused in a recent magazine article, according to which the star's magnetic field clears out the innermost part of the protoplanetary disk. When the planet reaches the clear space, according to this theory, the disk's viscosity no longer affects it, so it stays put. Marcy doesn't react as positively as we'd expected. "That was the mechanism that was favored as of a month ago," he says with a laugh. "But this field is evolving so quickly, there's another mechanism that's come through the grapevine. Doug Lin and a couple of other astronomers have pointed out that a protoplanetary disk might behave like Saturn's rings, where you get moons gravitationally shepherding the rings and each other: they get locked into resonant orbits, like 2:1 or 2:3. Neptune and Pluto are the same: they're in a 2:3 resonance. You could have a string of planets with commensurate orbital periods, shepherding each other, sort of keeping each other at bay against inward migration."

The close-in Jupiters are not good candidates to harbor life, of course. Quite aside from the fact that they are most likely gas giants rather than solid-surfaced planets, the proximity to their stars implies temperatures much too high for liquid water to exist. But Marcy and Butler have discovered several planets that orbit much farther from their stars. One such planet belongs to "47 Ursae Majoris" (or "47 UMa") a Sun-like star near the bowl of the Big Dipper. It has a likely mass of about three times Jupiter's mass, and it orbits at a distance from its star (2.1 AU) that in our own solar system would fall between the orbits of Mars and Jupiter. It could be called a "classical Jovian planet." Marcy and Butler also suspect that the star 55 Cancri A, besides having a close-in Jupiter, has a second planet: this one has a mass of about five Jupiters and lies, like 47 UMa's planet, at a distance corresponding to somewhere between Mars and Jupiter in our own solar system. Yet a third classical Jovian planet was discovered by the Mayor-Queloz group in 1998: it orbits a K-dwarf star named "14 Herculis." And in 1999, both Marcy's group and the Harvard-Smithsonian group announced the detection of two classical Jovian planets orbiting Upsilon Andromedae, a star that was already known to possess a "close-in Jupiter." Thus Upsilon Andromedae has at least three planets.

When they announced 47 UMa's planet in 1996, Marcy and Butler mentioned the possibility that it might possess liquid water, a possibility that seemed to make it more hospitable to life than the close-in Jupiters. In reality, given that it is probably a gas giant, the chances of finding bodies of liquid water there seem remote. But Jim Kasting, whose work on habitable zones was mentioned in Chapter 3, has come up with a new wrinkle on the matter. His graduate student Darren Williams raised the question of whether a moon orbiting 47 UMa's planet might be habitable. Together with Richard Wade, they did calculations suggesting that a moon orbiting the planet might possess an atmosphere, a rocky surface, oceans, and all other conditions necessary for life.[8] No such moon has actually been detected, of course. But if giant planets collect their own accretion disks as they form—and Jupiter undoubtedly did so, given the properties of its four large moons—it may be a good bet that all planets that large possess satellites of some kind.

Kasting, who played such a key role in expanding the habitable zones around stars, now finds himself under attack from the other direction—for setting the limits of habitability too narrowly. This change has happened because of the increasing belief in the existence of subsurface liquid water on Mars and Europa—water that is kept liquid by geothermal energy rather than by sunlight.

We get the chance to ask him about this when we run into him at a NASA astrobiology workshop. "If you define the habitable zone as every place that you can think of where you could get liquid water," he tells us, "then the whole solar system is a candidate. But where you can get liquid water on the surface of a planet has a totally different implication, which is relevant to the detection of life on extrasolar planets. The way we are going to detect life on those planets is by detecting the modification of the planet's atmosphere by living organisms—the presence of ozone particularly, which tells you that there's oxygen. Oxygen comes from photosynthesis, and for photosynthesis you need life at the surface, or very near the surface. With the right instruments, you could tell that Earth is inhabited from 10 parsecs away. But subsurface life on Mars or Europa—you'll never have a clue about it." Kasting also stresses (as does Chris McKay) how much more energy is available from sunlight (or starlight) than from geothermal sources. "It's a factor of 10^5 difference," he says, "I doubt that geothermal heat by itself can sustain liquid water on the surface anywhere, nor that it could provide enough energy to allow for modification of a planet's atmosphere on a scale that could be detected remotely."

If Kasting is going to make detectability the criterion for judging which habitats for life are worth thinking about, his postulated moon around 47 UMa's planet fails the test. He himself admits that there may not be even a theoretically possible method to detect the moon. When we ask him about this, however, he comes up with a long-shot method for detecting life there, even without detecting the moon itself. "If you're running a search for extraterrestrial intelligence, you could point your radio telescope at it," he says. "That's the only way I can see."

At the same time as they announced the planet around 47 UMa, Marcy and Butler announced another planet, this one around another Sun-like star named "70 Virginis." This announcement provoked tremendous controversy, however—not because of doubts about the planet's existence, but because of doubts as to whether it should be considered a planet or a type of very low-mass star called a "brown dwarf."

"Brown dwarfs" were so named in 1975 by a Berkeley graduate student, Jill Tarter, who is now a leading light in the search for extraterrestrial intelligence (see Chapter 7). At that time, they were merely a concept, and the concept had been developed largely by an Indian-born astronomer, Shiv Kumar, a decade earlier. Kumar knew that main-sequence stars come in a range of masses, which are identified by a difficult-to-memorize sequence of letters: O for the most massive, then B, A, F, G, K, and M. The Sun is a "G star." Kumar wondered whether "M stars"—dim "red-dwarf" stars with less than 60 percent of the mass of the Sun—were really the end of the line, or whether stars could form from even smaller chunks of a molecular cloud. He concluded that they could, but stars with masses less than about 8 percent that of the Sun would never achieve sufficiently high core temperatures to commence proton-proton fusion, the primary power source for main-sequence stars. After a brief period of deuterium (hydrogen-2) fusion, the star would collapse and just sit there, radiating in the infrared and barely, or not at all, in visible light. It would certainly not look "brown," but the name stuck.

There have been many claimed sightings of brown dwarfs over the years, some of which have been disproved, while others hover in a kind of astronomical limbo, neither damned nor blessed. But in 1996, a group at Caltech, led by Shrinivas Kulkarni, found an example that everyone accepts. It is a faint companion to an M-class star known as "Gliese 229," and therefore has been named "Gliese 229 B." It emits almost entirely infrared radiation. It lies about 45 astronomical units from its larger companion, and it has a mass about 0.5 percent that of the Sun, or fifty times that of Jupiter. Most interestingly, Kulkarni's group recognized the absorption lines of methane in the brown dwarf's infrared spectrum. Because methane would be destroyed at only a few hundred degrees Celsius, this finding confirmed the very low temperature of the brown dwarf's surface. Since the discovery of Gliese 229 B, infrared surveys have revealed dozens of extremely faint and cool dwarf stars—now called L dwarfs—scattered randomly throughout the solar neighborhood. Astronomers hope that these discoveries mark the beginning of a complete census of sub-stellar objects with masses intermediate between those of M dwarfs and planets.

The "planet" reported by Marcy and Butler around 70 Vir has approximately 9 Jupiter masses. (Remember that the radial-velocity technique cannot pin down the mass precisely, because the angle of view is not known). That's enough, in some astronomers' eyes, to bring it into brown-dwarf territory. It also has a highly eccentric orbit, which is also thought by some to be a hallmark of a brown dwarf. That's because a brown dwarf, falling out of the molecular cloud, would naturally take up such an orbit, whereas a planet, evolving within a stable circumstellar disk, might be more likely to have a near-circular orbit.

Marcy initially dismissed the brown-dwarf interpretation, but now he's more inclined to blur the distinction between the two kinds of bodies. Although they should have quite different origins—one condensing out of a molecular cloud, the other aggregating from smaller bodies in a disk—Marcy suggests that some bodies might have a kind of hybrid history. "You can imagine a scenario in which, out of the gas and dust that orbits a young star, the gas gradually condenses into an object, 5 or 10 Jupiter masses, without the benefit of grain-grain collisions and solid objects growing first. So it would be an object that grew in the disk like a planet, but not by the same process. Who knows, maybe even our Jupiter formed that way."

Marcy doesn't place much faith in the eccentricity of a body's orbit as saying anything about its status as brown dwarf or planet. "Pluto is a good example here in our own solar system: it's a planet in a highly eccentric orbit, probably because of past gravitational sling-shotting. It's conceivable that some fraction of the bona fide planets in our universe are residing in eccentric orbits because of these interactions. In fact, the galaxy could be full of planets that got ejected from their systems entirely. People have conjured up experiments to look for them."

The idea that highly eccentric orbits are a sign of brown-dwarf status has been deflated further by the discovery of planets with eccentric orbits but with seemingly low masses, well below anyone's minimum for brown dwarfs. In 1996, Marcy and Butler (and independently, a group at the University of Texas) reported finding a planet around the Sun-like star "16 Cygni B" that has a highly eccentric orbit but a minimum mass of only 1.5 Jupiter masses. In 1998, Marcy and Butler discovered a similar planet orbiting another Sun-like star, "HD 210277."

It's a real shortcoming of the radial-velocity technique that it cannot provide a definite measure of an extrasolar planet's mass, but only a lower limit. This, as mentioned earlier, is because the plane of the planet's orbit with respect to the line of sight is not known: a given radial-velocity signal could represent a relatively small planet in an orbital plane that we are viewing edge-on, or a much more massive planet or brown dwarf in an orbital plane that we are viewing nearly face-on. Recently, however, the true mass of one of Marcy and Butler's planets—the one that orbits 55 Cancri—was determined through a fortunate circumstance.

In 1996, a team led by Carsten Dominik of the Leiden Observatory in the Netherlands observed 55 Cancri, with the European Space Agency's Infrared Space Observatory, and detected an excess of infrared radiation coming from the star.[9] The team concluded that a disk of dust and debris, probably corresponding to our own Kuiper belt, was orbiting the star. Shortly after, David Trilling and Robert Brown of the University of Arizona imaged the disk directly using NASA's Infrared Telescope Facility on Mauna Kea.[10] They found that the disk was tilted 27 degrees from the plane of the sky. Making the reasonable assumption that the unseen planet is orbiting 55 Cancri in the same plane as the disk, and using Marcy's measurement of the amplitude of the radial-velocity signal, Trilling and Brown calculated the planet's mass: 1.9 Jupiter masses. This, of course, is well below brown-dwarf territory. Trilling and Brown expect that most or all stars with planets will also possess Kuiper-belt disks. Thus, it may eventually be possible to get a precise estimate of the masses of many of the extrasolar planets detected with the radial-velocity technique, including those whose status as planet or brown dwarf is controversial.

If the presence of planets implies a Kuiper-belt disk, does a Kuiper-belt disk imply the presence of planets? If so, identifying such disks could be a relatively simple way to home in on stars with planets. This kind of approach was used in 1998 by a group at the Joint Astronomy Center in Hawaii, led by Jane Greaves.[11] Greaves and her colleagues examined "Epsilon Eridani," the Sun-like star that is closest to our own solar system—it's just 10 light-years away. Marcy and Butler have observed this star for years with the radial-velocity technique, without seeing any hint of a planet. But Greaves and her colleagues, using the James Clerk Maxwell telescope on Mauna Kea, obtained a remarkable image: around the star lies a doughnutlike disk, surrounding a central hole, and one part of the doughnut is highlighted by an intensely bright spot. This spot, Greaves thinks, is dust that has either collected around a planet or is perturbed by a planet orbiting closer to the star, within the central hole. Marcy and Butler's negative results do nothing to rule this hypothesis out, of course. The planet might be too small or too far from the star to produce a detectable radial-velocity signal, or we may be viewing the disk-planet system so nearly face-on that even a large planet causes little motion of the star along the line of sight.

Like Queloz and Mayor, Marcy and Butler have expanded their search to cover a much larger selection of stars. Among the stars they are now monitoring are fifty "M dwarfs"—the lowest-mass main-sequence stars. In mid-1988, Marcy and Butler announced the detection of the first planet orbiting one of these M dwarfs—a star called "Gliese 876" that is only 15 light-years from earth. (A group at the Geneva Observatory in Switzerland, led by Xavier Delfosse, independently detected the same planet.) The planet is another close-in Jupiter, but its orbit is more elliptical than the others.

M dwarfs have only about 1 percent of the luminosity of the Sun, so life on a planet at 1 astronomical unit (the Earth's distance from the Sun) would seem impossible. And if one moved a planet in close enough to have liquid water, the planet would become tidally locked. The conventional wisdom is that such locking would make life impossible because the atmosphere would freeze out on the dark side of the planet and you'd end up with no atmosphere at all.

Marcy doesn't buy that. "Once you bring a planet close enough that the tidal issues come into the picture, suddenly you can draw upon the tidal energy, like the tidal flexing of Io or Europa, and then you realize that you don't need the luminosity of the star at all to heat up the planet. There could be not only subsurface liquid water, but also surface liquid water warmed by this tidal effect. It doesn't matter what luminosity the star is, if you bring the planet in close enough. You could have the conditions for life, maybe without photosynthesis." This scenario refers to a terrestrial planet, of course. Marcy's planet orbiting Gliese 876 is presumably a gas giant like Jupiter, and thus is not likely to possess a surface that could hold bodies of liquid water.

The main attraction of M-type stars is that they are so common. "Of the stars within 20 light-years of Earth," says Marcy, "80 percent are M-type stars. The missions that NASA is gearing up to carry out in the next three decades are all going to be most sensitive to the nearby planets. So that's why we're monitoring M-type stars: we are specifically trying to do reconnaissance for the later NASA missions. Maybe we will find the Jupiters and Saturns, and hopefully NASA will find the Earths."

The reason that Marcy doesn't expect to find the "Earths" himself is that the spectroscopic technique seems to be up against a limit of sensitivity. Right now, Marcy and Butler's instrument can resolve velocity differences down to about 3 meters per second—about the size of the wobble induced in our own Sun by the presence of Saturn. The Earth induces a wobble of only about 9 centimeters per second. Such a tiny wobble, Marcy believes, would be blurred out by a "fluttering" of the star's apparent velocity, caused by the roiling motions of gases on the star's surface. As it is, Marcy is compelled to select stars that rotate very slowly, for these have the calmest surfaces.

Among the methods that NASA is considering to extend the search for extrasolar planets, one involves the same method as was used to detect

Algol's stellar companion in the last century—the observation of the dimming of a star as an object passes in front of it. A proposed mission that uses this technique is the Kepler mission, developed by scientists at Ames Research Center, led by William Borucki.[12] This is a space-borne photometric telescope—a telescope able to measure the amount of light coming from each patch of sky with great precision. The plan is to put it into orbit around the Sun at a considerable distance from Earth. Once in space, the telescope will stare unblinkingly at a dense star field in the constellation Cygnus, not far from the bright star Vega. Over the course of 4 years, it will continuously monitor the brightness of 80,000 stars within its field of view, waiting for one of them to dim slightly for a few hours. If one does, the mission's Earth-based computers will merely take note of the fact. But if the same star dims again later, the computers will form the hypothesis that the two dimmings were caused by the passage of a planet in front of the star—a planet whose orbital period corresponds to the time interval between the dimmings. Then, the main test will come when the third dimming falls due: If it happens as predicted, the presence of the orbiting body will have been confirmed.

Although the chances that any particular star's planets pass directly between the star and Earth are quite low, the huge number of stars that Kepler can monitor encourages the mission's proponents to believe that it will detect hundreds of planets during its 4-year mission. Specifically, they predict that Kepler will detect about 480 terrestrial-type planets, 160 close-in giant planets, and 24 giant planets in outer orbits. Of course, there are some untested assumptions involved in this prediction, including the assumption that terrestrial planets are common.

The photometric method gives not only the time of the dimming, but also its duration and its depth (i.e., the fractional decrease in brightness). This information, combined with the known characteristics of the star, allows the planet's size and its likely temperature to be determined. If a particular giant planet were detected both with the photometric method and the radial-velocity method, its mass and size would both be known, allowing its density to be computed. This in turn would allow for informed guesses about its composition. Thus, from what seem like very impoverished data—three blips in a photometric record and some oscillations in a spectrum—astronomers may be able to draw surprisingly detailed conclusions about a planet's habitability.

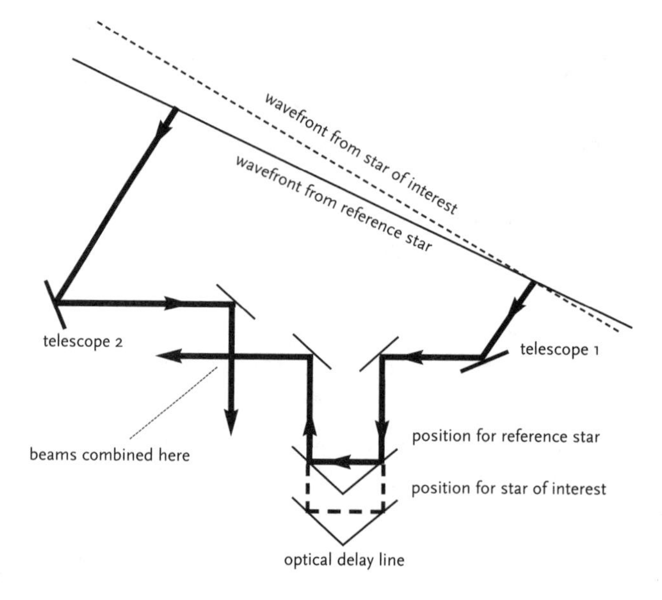

5.2 Interferometry using a reference star. Two beams of light from the reference star, captured by the two telescopes, are combined to produce interference fringes. To do the same for the star of interest, whose light waves are approaching the telescopes at a slightly different angle, the length of the optical delay line has to be adjusted slightly. The amount of this adjustment gives the angular separation between the two stars.

The Kepler mission is competing (unsuccessfully, so far) for a slot in the so-called Discovery Program, a series of NASA missions characterized by the slogan "Faster, cheaper, better." These missions are supposed to cost less than $300 million and to take no more than 3 years to complete. The Discovery Program is a response to budgetary constraints and to the megaprojects of the past, some of which became so overloaded with bells and whistles that their failure was all but assured.

There is another planet-finding technique, however, into which NASA is planning to invest most of its efforts, and that is optical interferometry, which combines the light of two or more telescopes that are located at some distance from each other, thereby gaining the resolution of a single telescope with a diameter equal to the distance between the two. The method will be used not only to detect the side-to-side wobble of a star that is orbited by planets, but eventually to image those planets directly and even to analyze their atmospheres.

The tricky part is combining the light from the two telescopes. This must be done in such a way as to preserve information about the phase relationships (the relative positions of the individual wave fronts) of the light reaching them. Thus, it is necessary to measure the length of the paths taken by the starlight entering the two telescopes, with a level of precision down in the billionths of a meter.

To explore and develop the techniques of optical interferometry, NASA has constructed the Palomar Testbed Interferometer (PTI) (see Color Plate II).[13] PTI is run by a large group of JPL scientists, headed by Mike Shao; but when, sometime after midnight, we climb down from the catwalk of the Hale telescope and tap on the door of the PTI's "beam-combining building," we are greeted by a lowly Caltech undergraduate, Ben Lane. Though he appears altogether too young to be entrusted with such an expensive piece of hardware, Ben is intimately familiar with all its mysteries, because he helped assemble and test it. He is soon snowing us with an avalanche of techno-speak—"dichroics," "delay lines," "tilt tables," "heterodyne lasers," "single-mode fibers," "fringe trackers," and so forth. Luckily one of us (Koerner) uses interferometry (at radio wavelengths) in his own research, so at least the principle of the thing is clear.

Basically, two telescopes (the "north" and "south" telescopes, on this particular night) look at the same small region of sky, which is chosen to contain at least two stars: the star of interest and a nearby bright reference star. The reference star is taken as a fixed landmark; the star of interest, on the other hand, will be changing its position slightly from night to night, if it possesses planets. Thus, the aim is to determine as precisely as possible the position of the star of interest with respect to the reference star. (The reference star may have some "wobble" of its own, of course. Therefore, in practice it's necessary to use more than one reference star, in order to determine which motions are intrinsic to the star of interest.)

Because the reference star and the star of interest are at slightly different positions in the sky, the wave fronts of the light coming from them approach the telescopes at slightly different angles. For example, if the peaks of the waves from the reference star happen to strike the two telescopes synchronously (in phase), then the wave fronts from the star of interest will strike the two telescopes slightly out of phase—the peaks may reach one telescope at the same time as the slopes or the troughs of the waves reach the other telescope. The interferometer has to measure the phase difference between waves from the two stars: this is a measure of the angular separation between them, as measured along the axis running through the two telescopes.

To accomplish this, beam splitters inside each of the telescopes separate the light from the two stars and send the resulting beams down evacuated pipes into the combining building. Thus, a total of four "starbeams" enter the building. Once inside, the two beams from the north telescope are sent on an out-and-back trip to a mirror that is perched on top of a miniature railcar. By adjusting the position of the railcar on its track, the total length of this "optical delay line" can be adjusted. The beams from the other telescope are not delayed.

Next, the reference-star beams from the north and south telescopes are brought together, and the delay line is adjusted until the wave fronts from the two beams are in register, producing a characteristic pattern of interference "fringes," like the colored bands of light on an oil slick.

Finally the two beams from the star of interest are combined. These beams will not produce the same pattern of fringes as did the two reference-star beams, since they were in a different phase relationship when they entered the two telescopes. To produce the same pattern of fringes, the length of the delay line must be changed slightly. And we mean slightly—a few billionths of a meter. This is done, not by moving the railcar, but by use of tiny piezoelectric actuators behind the mirror. The difference in delay-line length, which is measured with laser beams that traverse the entire optical pathway, provides a measure of the angular separation between the two stars. It's only a one-dimensional measure, of course. To get a two-dimensional measure, a different pair of telescopes (say, "north" and "east") can be used, or one can simply wait for the Earth's rotation to twist the original pair of telescopes into a different orientation with respect to the stars.

Sounds complicated? Possibly, but you are using the same technology yourself all the time, without even thinking about it. When you sense that a sound is coming from a particular direction, that's in part because your brain is performing interferometry on the sound waves reaching your left and right ears. To put it more precisely, your brain converts the sound waves into patterns of neural impulses and performs interferometry on those patterns. In fact, your brain goes one better than PTI: instead of making trial-and-error adjustments in a single auditory delay line, it simply feeds the signals into a massively parallel set of lines representing all possible delays. If Ben Lane could be persuaded to set up the PTI in the same way, results would be nearly instantaneous.

PTI currently has an astrometric precision of about 100 microarcseconds, and it will ultimately have a precision of about 50 microarcseconds, which means that it should be able to detect a nearby star's wobble caused by the presence of a Jupiter-like planet.

PTI can also operate in a simpler mode, in which it simply looks at a single star through the two telescopes. This technique cannot detect a star's wobble, but it can determine whether a star's image is a "point source" or is something more complicated, such as a binary pair. In 1997, Mike Shao's group used this method on 51 Peg, the star whose now almost sacred "planet" was detected by Queloz and Mayor. Shao's group claimed to find that 51 Peg was a binary system, meaning that the "planet" was not a planet at all but a small companion star. They wrote an abstract on this finding that got onto the Internet and then into the pages of *Science*. But follow-up work showed that this result was incorrect: 51 Peg is indeed a single star, and the planet exists after all. When, on another occasion, we have the chance to ask Shao about this, he laughs and says "The final word is, one of our team members was a little bit overly enthusiastic. Once we got rid of two large sources of error, the wiggles weren't there." Since Queloz is now a member of Shao's team, the retraction has presumably smoothed the atmosphere within his lab.

PTI's main function, however, is not to detect or refute planets but to serve as a prototype for a succession of bigger and better interferometers. First, the Keck telescopes will be set up in differential interferometric mode: this will involve not just the integration of the two existing 10-meter telescopes but the construction of four small "outrigger" telescopes that will all be optically connected.[14] The Keck Interferometer is expected to have a precision of 10 to 30 microarcseconds and to be capable of detecting the wobble caused by Uranus-like planets orbiting stars as distant as 20 light-years from Earth.

The Keck telescopes will also be capable of operation in another mode of interferometry, which should permit direct detection of the close-in Jupiters. In this mode, two starbeams from the star of interest are combined, and the fringe pattern measured. The measurement is done first at visible wavelengths—a region of the spectrum in which the star so greatly outshines the planet that the latter makes essentially no contribution to the fringe pattern. Then the measurement is repeated at infrared wavelengths—a region in which the planet, because it is heated by the nearby star, does make a small but detectable contribution. Because the planet is at a slight distance from the star, its contribution is out of phase, and it therefore shifts the overall fringe pattern slightly. By measuring this phase shift at many different orientations, an image of the star and its planet can be reconstructed. Thus, Shao hopes to be able to confirm Marcy and Butler's planets and to detect other close-in Jupiters.

The next step is the launch (planned for the year 2005) of the Space Interferometer Mission (SIM), a set of six small telescopes arrayed along a 10-meter truss.[15] Operating in the differential astrometric mode, SIM will have a precision of 1 microarcsecond—corresponding approximately to the distance between the left and right shoulders of an astronaut on Mars, as seen from Earth! This is sufficient to detect (or rule out) Earth-like planets around the twenty or so nearest stars.

Then, in the following decade, if all goes well, will come the Terrestrial Planet Finder TPF (see Color Plate 12). This infrared interferometer is expected to be about 100 meters long—an order of magnitude larger than SIM. It will operate in yet another mode of interferometry, called the "nulling mode," originally thought up by Stanford physicist Ronald Bracewell.[16] In this technique, two starbeams from a single star are combined in such a way that the peaks of the waves in one beam line up with the troughs in the other. This causes "destructive interference," greatly reducing the measured intensity of the light from the star. Extending outward from the image of the star, however, are a set of parallel bands (fringes) in which light from objects such as planets is alternately suppressed and enhanced. By rotating the entire interferometer around the line of sight, the image of the planet is made to blink on and off as it moves in and out of these bands. Thus, by making measurements at all orientations, TPF will be able to reconstruct a direct two-dimensional image of the star and all its orbiting planets.

A major potential problem with this approach may be presented by dust. The inner parts of our own solar system contain enough dust to outshine the Earth at infrared wavelengths. (The scattering of sunlight by this dust produces a glow in the night sky known as the "zodiacal light.") The obscuring effects of the dust can be avoided by sending TPF out to the region of Jupiter, where viewing conditions are much better. But what if the target stars and their inner planets are also embedded in cocoons of dust? To investigate these "exozodi" (a word that only Scrabble players could love), they will be imaged with the Keck telescopes, operated in the nulling interferometric mode. If exozodi are found to be especially common or dense, the telescopes on TPF may have to be made much larger than originally planned.

Because of its ability to directly image planets, TPF will be able to obtain spectra of the infrared light coming from the planets. Within the infrared region lie telltale absorption bands for carbon dioxide (indicating an atmosphere), water (suggesting that the planet may be habitable), and ozone. Ozone is an indicator for the presence of oxygen, which on our planet at least, was created by living creatures. Thus, TPF not only may tell us where the terrestrial planets are but also may give us strong hints as to which planets actually harbor life.

Although the TPF (and a rival European device, the Infrared Space Interferometer) lie at least 10 years in the future, Shao is busy thinking about what might come after it: "You probably haven't met Dan Goldin [the NASA Administrator]. He likes long-term visionary goals. So we have formed a small group to look at ways of making major advances over SIM or TPF. One of the things we're asking is: What would it take to produce a multipixel image of an extrasolar planet? One person at Lawrence Livermore has come up with the concept of building a very large aperture telescope out of a thin film of plastic, a Fresnel-lens concept, like those things you see stuck on the back windows of minivans. Yes, they make very poor images. But if you can measure all the aberrations, you can manufacture a corrector to take them out. There may be ten million points where you need to correct, and the way you do that is to make the corrector with semiconductor manufacturing techniques. You can get to an image that is diffraction limited [i.e., as close to perfect as you can ever get]. What we're looking at now is a 25-by-25 pixel image, where you see the continents and oceans. That would require twenty-five telescopes, each 40 meters in diameter, spread out across maybe 400 or 500 kilometers of space. Our feeling is that flying these things 400 kilometers apart is not that hard. But getting the 40-meter telescopes at a price the country is willing to pay for them, that's by far the hardest challenge. It may take one major invention beyond the plastic film."

Even supposing that habitable planets are commonplace, and that a Life actually arises on many of them, we are still left with great uncertainty about the course such a Life might take—about evolution, in other words. In studying evolution, we presently have only terrestrial evolution to go by. If terrestrial evolution has been simply a concatenation of chance events, leading to a completely arbitrary set of organisms, then we can say little about what we might find elsewhere in the galaxy, except that it will likely be very alien to us. If, on the other hand, terrestrial evolution has been guided by universal principles, then there should be some basic similarities between what we see on Earth and what we may find elsewhere. How far might such similarities go? Would they extend, for example, to the appearance of organisms with intelligence, curiosity, and a moral sense like our own? These are the questions we explore in the next chapter.

SUGGESTED READING

Marcy, G., and P. Butler. "Discovery of Extrasolar Planets" [available at http://www.physics.sfsu.edu/~gmarcy/planetsearch/planetsearch.html].

Angel, J.R.P., and N.J. Woolf. "Searching for Life on Other Planets." *Scientific American*, April 1996, 60–66.

Croswell, K. *Planet Quest: The Epic Discovery of Alien Solar Systems.* New York: Free Press, 1997.

Schneider, J. "The Extrasolar Planets Encyclopedia" [available at http://www.obspm.fr/encyl/encycl.html].

Penny, A.J. "Extra-Solar Planets" (a section of the website for the European Space Agency's "Darwin" space interferometer mission). [available at http://ast.star.rl.ac.uk/darwin/planets/welcome.html.]

6

What Happens in Evolution?

Chance and Necessity in the Origin of Biological Complexity

The Santa Fe style is "faux adobe." The parking garage where we leave our car, for example, is held together by steel girders, but they emerge to the exterior transformed, as if by the glare of the New Mexico sun, into oaken roof beams. In the central plaza, whose every building has been retrofitted with a mud-colored façade, tourists pore over "Indian" blankets—blankets that are hot off the truck from a factory in Mexico. The city, in short, is a Disneyfied pueblo, but without a hint of Disney's self-mockery.

Santa Fe has been billed as a meeting place of cultures: the place where intellectual modernism, represented especially by the complexity theorists of the Santa Fe Institute, encounters the timeless, earthbound spirituality of the Anasazi. In the resulting ferment, we're told, the old symmetry breaking that froze human thought into narrow subdomains is reversed, and a unified view of Life and Mind is being forged. Such, at least, seems to be the message of books like George Johnson's *Fire in the Mind,* which attempts a seamless fusion of the two worldviews.[1]

The truth may be a little different. The real indigenous culture is aptly represented by a temporary show at the Institute of American

Indian Art, which includes a teepee made of empty liquor bottles and a wall covered with the scrawled confessions of pregnant teenagers. And the scientists have little time for the Immanent. Stuart Kauffman, the object of our visit, has his hands full with a start-up company called Bios Group, whose mission is to turn complexity theory into a cash cow.

Once at the Bios Group office, just a few steps from the central plaza, we are kept waiting 2 hours beyond our scheduled meeting time while, according to his assistant, Kauffman gets a massage. We are busy plotting the writer's revenge—character assassination—when the great man finally appears and invites us to lunch.

As it turns out, Kauffman is too interesting to sustain our animosity. His mind, like its subject matter, is complex: it operates near the boundary between ordered modesty and egocentric chaos. That shows itself, for example, in a tension that arises whenever the question of originality comes up. "I'm not the first one to think about this," he says at one point, "although I think I'm the first one to think about it the way I did. Melvin Calvin in his book on chemical evolution has the idea, so I'm certainly not the first one who's played with it because his book was in '69—I didn't get the idea from Calvin but this doesn't matter." Or: "Freeman Dyson's got the essential idea, but he makes a kind of argument in which, with full affection and respect for him—I think very highly of him—it's also true that he assumed his conclusions, he actually begs the question, if you read the book. And my reaction was, 'Gosh, I did this in 1971, and I like my idea better than I like these guys', and even if they're famous, I'm going to do it again.'"

In the rapidly evolving world of complexity theory, it's hard to keep track of who first thought of what. Made possible by the advent of powerful computers, the field is still in its "Cambrian explosion" phase, when all kinds of exotic notions make a brief appearance on the stage, only to succumb to Chance or Necessity. Kauffman himself has put forward theories on all kinds of topics: how life originated, how multicellular organisms function, how biological evolution works, and how human institutions develop. And though he wrote in his 1995 book *At Home in the Universe*, "I am not brazen enough to think about cosmic evolution," he now does that too.

What unifies Kauffman's thought is a single leading idea: complex systems develop, in part at least, according to intrinsic rules. Something complex, such as a human being, didn't necessarily get that way because of an equally complex description, instruction set, or sequence of causes—but spontaneously. "Order for free," as Kauffman likes to say.

If this is true, one might expect there to be underlying similarities between different kinds of complex systems. That notion, indeed, has been kicking around for centuries—it's implicit in metaphorical expressions like the "body politic," for example. But the self-appointed task of the Santa Fe theorists is to give formal mathematical clothing to that bare idea: to establish laws of complexity. In that sense, the theorists are searching for deep principles that could underlie the origin and evolution of living systems wherever they might be found.

As a simple example, Kauffman cites the work of physicist Per Bak and his colleagues, who computer-modeled the formation of a pile of sand grains on a table.[2] As the grains rain singly and steadily down from above, the pile gradually steepens until it reaches a critical state. Then, as more grains are added, a series of avalanches begin, some small and some large. The size of the avalanches follows a power-law distribution, with many small avalanches and fewer large ones. A given grain of sand, when added to the pile, can unleash a small avalanche, or a large one, or do nothing. Thus, a very simple repetitive input drives the system to a state where it shows complex but mathematically describable behavior. "Self-organized criticality," Bak calls it.

Right away, one of the main problems of complexity theory raises its head: Does it have anything to do with reality? Do real sand grains, poured onto a real table, also unleash a power-law distribution of avalanches? "There's two groups working on this," says Kauffman, "and one says yes and one says no. People also do it with rice, and it turns out that it works with short-grain rice but not with long-grain rice. My own view of this is that Per Bak did an absolutely brilliant piece of work, and I think that there's a good chance he captured something very general; if it's wrong in details here and there it's still extremely interesting."[3]

Kauffman's work on the origin of life offers a very different vision than that of the San Diego researchers such as Leslie Orgel and Jerry Joyce. "In 1971, I was wondering about the origin of life," he says, "and everyone was thinking 'RNA,' and I found myself thinking 'It's just too special.' Suppose the laws of chemistry were just a little bit different, so that you couldn't get the beautiful double helix of DNA and RNA—do you really think that you wouldn't get life?"

Kauffman's approach was to ask whether it was possible to obtain, in a computer at least, a diverse set of molecules that interacted with each other to produce a self-sustaining system: an "autocatalytic set."[4] In such a set, no individual type of molecule is required to reproduce itself, since molecule A may be produced through the efforts of molecules B and C, molecule B through D and A, and so on. So RNA, with its potential for self-replication through the base-pairing mechanism, loses its special allure. Kauffman favors polypeptides (chains of amino acids, probably shorter than those forming the proteins we're familiar with) as the main chemical agents in the set. Most enzymes in the contemporary world are made of protein rather than RNA, after all, which suggests to Kauffman that they have a higher probability of acting as catalysts than RNA molecules. Perhaps RNA was an afterthought of a peptide world, just as Orgel has envisaged DNA as an afterthought of the RNA world. Recently, Claudia Huber and Günther Wächtershäuser, working at the Technische Universität of Munich, have provided some supportive evidence for this notion: they demonstrated for the first time that a variety of very short peptides can form from amino acids on the surface of metal sulfide grains, in conditions resembling those found at deep-sea vents.[5]

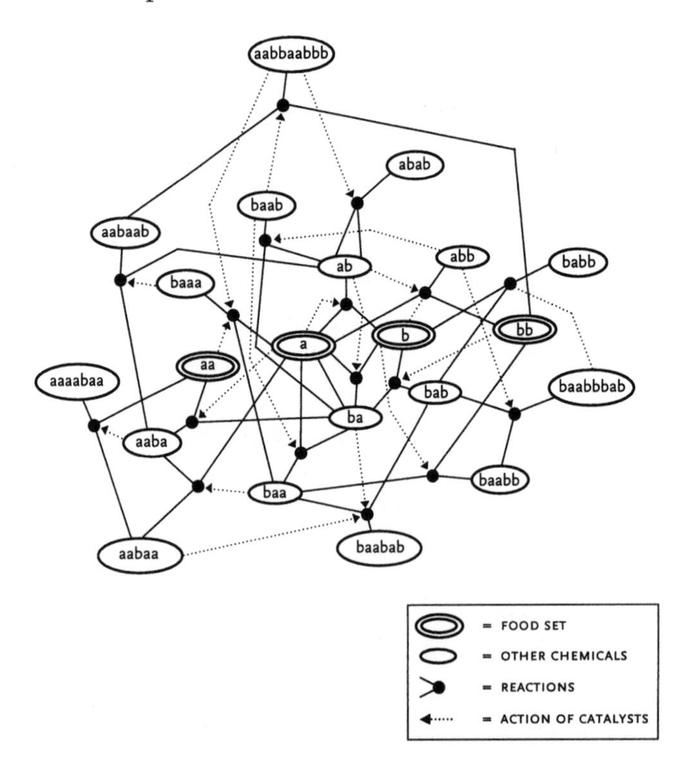

6.1 An autocatalytic set. Food molecules (a, b, aa, bb) are built up into a self-sustaining network of polymers, some of which catalyze reactions within the network (courtesy of Stuart Kauffman).

The property of autocatalysis does not emerge gradually, as Kauffman adds more and more diverse molecules to his soup. Rather, there is a sudden "phase transition" in which the whole system gels into a large, interconnected, self-sustaining network. Beyond that point, the system maintains its own complex order indefinitely, while growing and perhaps dividing into subsystems. It is not a closed system, however, in a thermodynamic sense: energy has to be constantly supplied from the outside in the form of "food" molecules.

A critical question, of course, is how many different molecules are required to obtain an autocatalytic set? That depends on the probability that a randomly selected molecule will catalyze a given reaction. And that in turn depends in part on how long a polymer has to be before it makes a decent catalyst. The longer it has to be, the greater the number of rival polymers that fail to do the required job. Considering one particular catalyst—the ribozyme known as RNA replicase that is envisaged to lie at the heart of the RNA world, Jerry Joyce and Leslie Orgel have voiced the opinion that it must be at least fifty bases long. There are 10^{30} possible RNA sequences of this length. If we are in general faced with 1-in-10^{30} probabilities, we might as well forget about spontaneously arising autocatalytic sets.

Kauffman, however, places the odds much higher, for three reasons: (1) he doesn't think polymers have to be that long, (2) he thinks that very large numbers of different polymers can catalyze the same reaction, and (3) he thinks that polypeptides make better catalysts than RNA molecules. He points to work on the binding of antibodies to antigens, which suggests that a randomly chosen antibody has about a 1-in-10^5 chance of binding to a particular antigen. Although catalysis may require more than just binding, Kauffman thinks that the probability of a randomly chosen polypeptide catalyzing a given reaction may be in the range of 1 in 10^9 or 10^{10}. If it's 1 in 10^9, Kauffman calculates that a set of molecules will become autocatalytic when it contains a mere twenty thousand different molecular species. "That's trivial," he says.

Orgel and Joyce aren't too impressed. "It's mostly nonsense," says Orgel in his usual downbeat way. "And the reason that it's nonsense is not that there's anything wrong with the mathematics, but because it assumes that the molecules have properties that they're not likely to have. Assuming that a peptide of twelve or so amino acids is going to be a rather specific catalyst is excessively unlikely. There are experiments on this."

Joyce is more equivocal. "It's a big world," he says. "There's room for different points of view. In some special cases, a product will feed back earlier on in the loop, and you might get a little cycle of reactions until it runs out of input, and those are the kinds of things that Stuart talks about. That's fine. Then Stuart goes further though, and says it's all going to close and start spinning around, and you'll get complex cycles that branch and get ever more complex. That one I don't buy. I don't buy it because of problems to do with specificity, and the fact that molecules have to react in real time with kinetic rates, association rates, rates of chemical steps, and infidelity of chemical steps. So that's where I agree and where I don't."

To Kauffman, the clincher will come when autocatalytic sets are actually generated in the laboratory—something that he says several groups are working on. "I'm pretty persuaded that we'll get autocatalytic sets as an emergent phenomenon. Even if we do, though, and we've got a new Life started, none of our theories prove that that's how life actually started on Earth."

If metabolism by itself is complex, a whole other level of complexity exists at the level of macroscopic organisms such as ourselves. We are built of a host of different cell types, each carrying out its own highly specialized task. Nerve cells send electrical signals down axons, pancreatic islet cells churn out insulin, muscle cells contract, and so on. Yet, with some exceptions, all cell types in our body contain the same suite of genetic instructions as the fertilized egg we arose from. As Dolly the sheep has shown, every cell in our body has the information required to build all the tissues of a complete individual. Yet, somehow, complex specialization is the rule. What is the source of that complexity?

To tackle this problem, Kauffman moved to a slightly more abstract level of analysis than that of autocatalytic sets. Whereas the autocatalytic sets are explicitly composed of chemical compounds, Kauffman has explored cellular specialization with logical systems known as Boolean networks. Such a network is composed of units that can be in one of two alternative states—"on" or "off." Kauffman often describes them as "lightbulbs." Whether a particular lightbulb is on or off is determined by the state ("on" or "off") of a certain number of neighbors in the network and by a set of logical rules. For example, "lightbulb A" might be connected to two neighbors (B and C), and the rules might be as follows: If B and C are both "on," A is turned "on"; if B is "on" and C is "off," A is turned "off"; and so on. The network goes through an it-

Now you don't see it—now you do.
Life under a quartzite pebble in Death Valley
(photos by the authors).

Life within rock. Above: University Valley, Antarctica. The brown sandstone outcrops in the foreground are colonized by microorganisms. Below: Cross section of the surface of the sandstone. The black and white zones are formed by endolithic fungi and algae, which together form a lichen. The green zone is formed by algae and cyanobacteria (photos courtesy of Imre Friedmann).

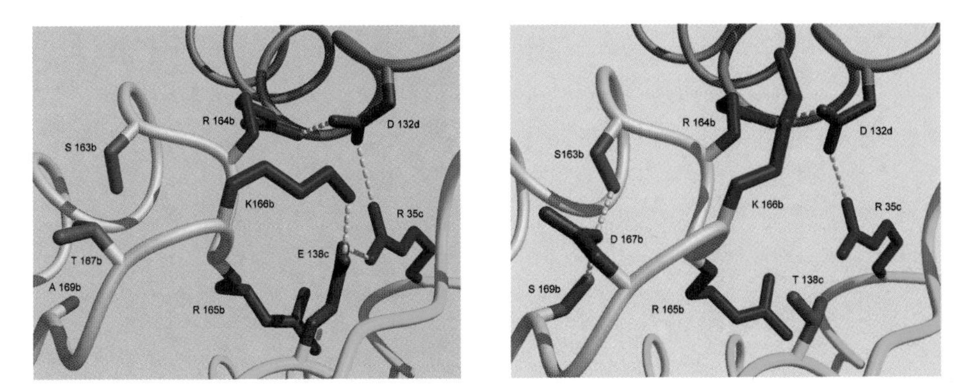

Coping with heat. One way organisms may survive at very high temperatures is by stabilizing their protein molecules with ion pairs. The diagrams compare the three-dimensional structure of part of a protein enzyme from a hyperthermophile (optimal temperature 100°C—left panel) with the equivalent region from a thermophile (optimal temperature 88°C—right panel). In the hyperthermophile, the structure of the protein is stabilized by ion pairs formed between the amino acid glutamate (E 138c) and three neighboring amino acids (R 35c, K 166b, and R 165b). In the thermophile, this glutamate is replaced by another amino acid, threonine (T 138c), which cannot form ion pairs, so the protein molecule is more flexible (courtesy of David Rice and Patrick Baker).

Artist's impression of the robotic exploration of Europa's ocean. The cryobot has penetrated the overlying ice and released the hydrobot, which is exploring a volcanic vent. The liquid-water ocean is probably far deeper than suggested here (JPL).

The Orion nebula, in a Hubble Space Telescope image (C. R. O'Dell and S. K. Wong, NASA).

COLOR PLATE 4

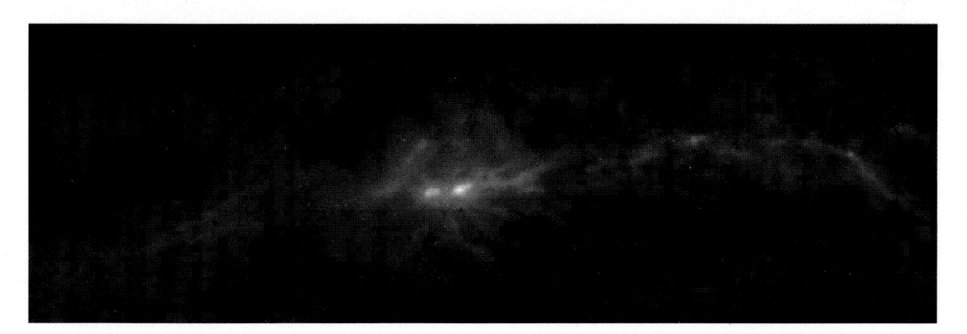

Starbirth. This is a map of far infrared radiation from dust in Orion. Some of the youngest "protostars" known appear as bright spots in the center. The image was obtained with the James Clark Maxwell Telescope on Mauna Kea, Hawaii, and is provided courtesy of Doug Johnstone (CITA) and John Bally (CASA).

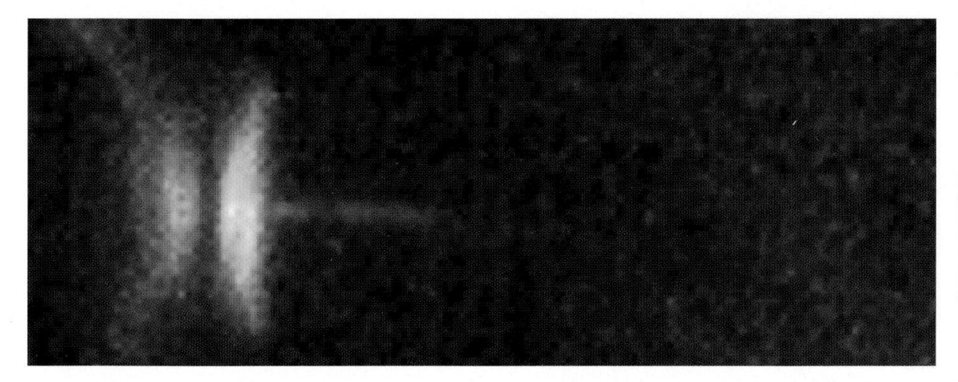

Protostellar object HH-30 in the Taurus-Auriga cloud. The star itself is hidden by the disk, but its light illuminates the upper and lower faces of the disk (bright vertical bars at left). An ionized jet (red) extends for billions of miles from the star along the disk's axis of rotation (Mark McCaughrean and C. R. O'Dell, NASA).

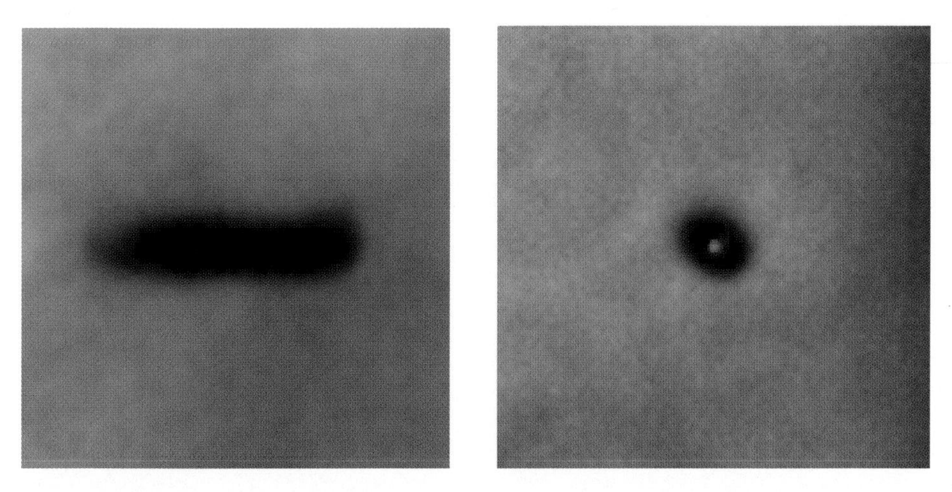

Protoplanetary disks or "proplyds" in the Orion nebula. In the left-hand image the disk is seen edge-on, silhouetted against a background of brightly lit clouds of gas and dust. The central star is obscured by its disk. In the right-hand the disk is seen almost face-on, and the central star is visible (Mark McCaughrean and C. R. O'Dell, NASA).

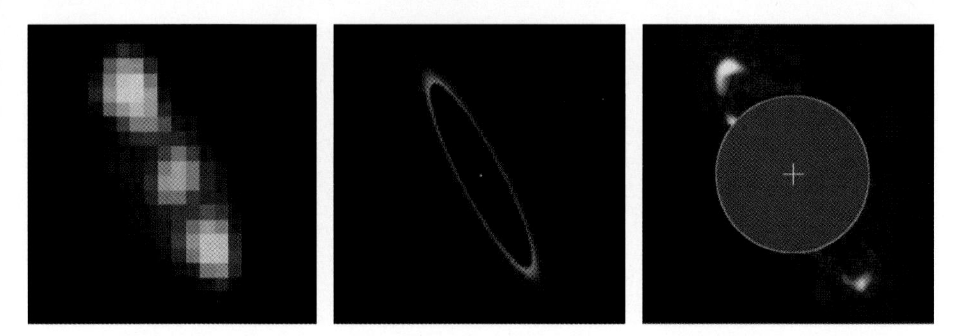

Planet birth? This is a young star, HR 4796A, imaged in mid-infrared light with the Keck Telescope (left panel) and in near-infrared light by the Hubble Space Telescope (right panel: a coronagraph blocks the light from the start itself). A disk of infrared-emitting dust orbits the star. Computer modeling of the distribution of infrared emission indicates that there is a hole in the disk near the star (center panel). The hole is about the diameter of our solar system, and is probably caused by the presence of rocky planetesimals or even newly accreted planets (left and center panels by David Koerner and colleagues; right panel by Brad Smith and colleagues, Institute for Astronomy, Hawaii).

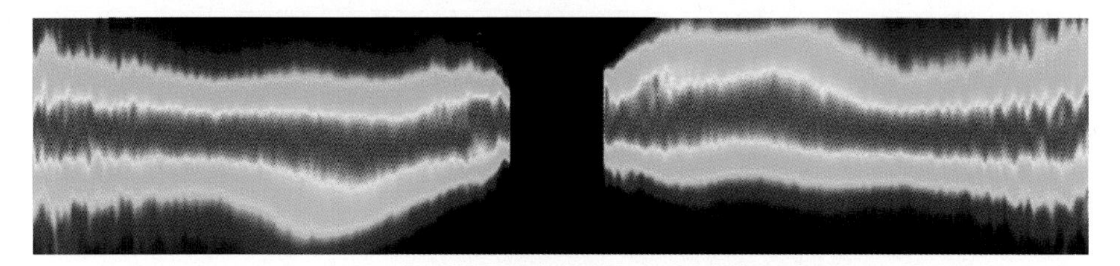

False-color image of the debris disk around Beta Pictoris, taken with the Imaging Spectrograph of the Hubble Space Telescope. The disk is seen edge-on; the light from the star itself has been blocked by a coronagraph (gap at center). Note that a portion of the disk near the star is warped, probably by one or more unseen planets (Sally Heap, GSFC/NASA).

The Palomar Testbed Interferometer. Small telescopes in the three outlying buildings send starlight along the pipes to the central beam-combining building. The elongated shed extending to the right from the central building houses the optical delay lines (photo by the authors).

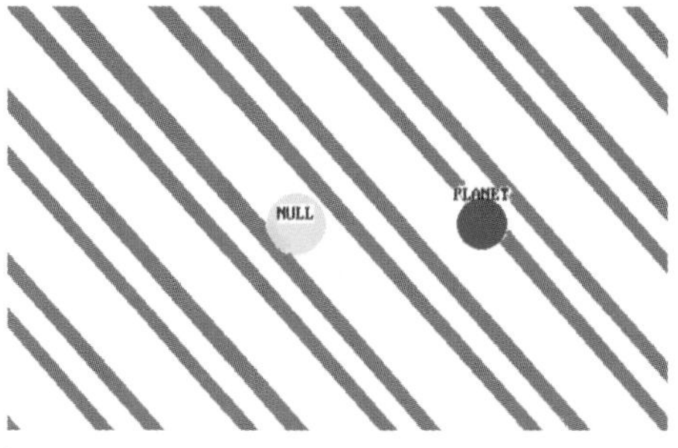

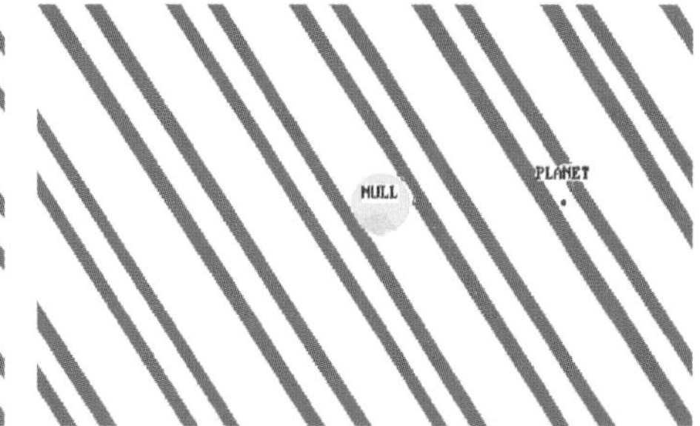

Imaging extrasolar planets by nulling interferometry. Top: Proposed design for the Terrestrial Planet Finder (JPL). Center: How TPF works. The interferometer creates bands of constructive (pink) and destructive (white) interference. A band of destructive interference is positioned over the star (yellow spot), thus blocking or "nulling" most of its light. As TPF rotates, the star remains nulled, but the planet (blue spot) blinks on (large spot, left) and off (small spot, right) as it is straddled alternately by constructive and destructive bands (courtesy of Nick Woolf). Bottom: From the sequence of blinks recorded during the rotation of TPF, an image of the star's planets can be computed. This is a simulated TPF image of the inner part of our own solar system. The three arrows represent the positions of (left to right) Venus, Earth, and Mars. Venus and Earth are easily detected, but Mars only poorly. The red spot farther out from the star is an artefact. Also, note that the interferometer generates two mirror images of each planet, on opposite sides of the star (JPL).

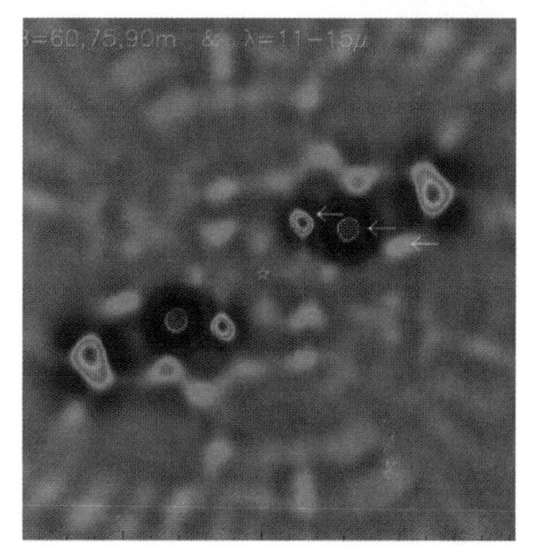

Creatures of the Burgess shale. A fearsome *Anomalocaris* (top center) has captured a trilobite. On the sea floor, from left to right, are a *Wiwaxia*, three *Hallucinogenia* (with spines protruding from their backs), a centipede-like *Aysheia*, and a five-eyed *Opabinia* (green). Above the latter are two *Marella* (courtesy of Simon Conway Morris).

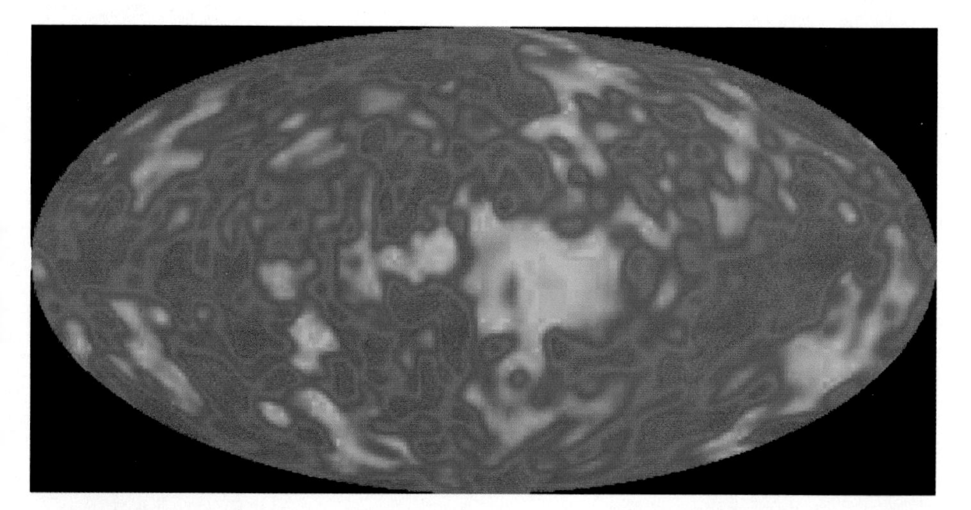

Ripples in the afterglow of the Big Bang. This is the cosmic microwave background, mapped by the COBE satellite. Red areas are parts of the sky where the background is very slightly warmer than average, blue areas are where the background is very slightly cooler. Temperature variations caused by the solar system's motion through space, and by the Milky Way galaxy, have been removed from the image (courtesy of Ned Wright, UCLA).

erative procedure in which, at each iteration, all the lightbulbs examine the states of the neighbors to which they are connected and then adjust their own states according to the prescribed rules.

As a Boolean network goes through iteration after iteration, it can show a variety of behaviors, depending on the number of units (N), the number of connections per unit (K), and the details of the connectivity rules. Because some state of the network must eventually be a repetition of an earlier state, the system must eventually enter a repeating cycle of states. But the length of the cycle can be as short as one iteration—meaning that the network simply remains in a fixed state—or as long as the entire number of different states that the system is capable of representing. With a reasonably large network, this number will be so huge that, over any reasonable period of observation, the system will seem to behave chaotically.

It is between these two extremes that interesting behavior patterns emerge. A well-behaved Boolean network has state cycles that are longer than one state but much shorter than the sequence of all possible states of the network. Because most possible states of the network are not represented in a given state cycle, the network may exhibit a large number of independent state cycles. However, there are many fewer state cycles than possible initial states. In other words, there is convergence: many different initial states lead eventually to states that belong to the same cycle. So the state cycles are referred to as "attractor," and the collection of states that feed into a single attractor is known as its "basin of attraction." That gives the attractors a kind of robustness: flipping a lightbulb or two often leaves the network in the same basin of attraction, so that it soon returns to the same attractor it was cycling through before the intervention. This robustness is a lifelike trait.

According to Kauffman, "well-behaved" networks are those in which K, the number of connections per unit, is small but greater than 1. With K = 2, for example, a Boolean network of N = 100,000 units will possess about the square root of 100,000 different state cycles (or about 316 cycles), and each cycle will also be about the square root of 100,000 states long (or about 316 states). This is surprisingly orderly behavior. It means that if you pick an initial state completely at random out of the $10^{30,000}$ possible states of the network, it will end up in one of 316 cycles, each only 316 states long. Thus, the vast majority of the system's possible states form the basins of attraction and are never visited once the system settles down.

With K greater than 2, networks tend to show chaotic behavior, but certain kinds of connectivity rules permit orderly behavior even with large K's. This is true if the rules are chosen so as to bias each lightbulb to a particular state ("on" or "off"), or if at least one of the set of lightbulbs connected to "lightbulb A" has one state ("on" or "off") that completely determines A's state (Kauffman calls this a "canalizing function").

What Kauffman observed was that, as the network enters a state cycle, a large sea of lightbulbs (something like 70 percent of the total in an N = 100,000, K = 2 network) goes into a unchanging state (either "on" or "off"). These "frozen" lightbulbs form an interconnected web. Scattered across this frozen sea are islands of "twinkling" lightbulbs that continue to change their state as the system progresses round and round the state cycle. The moment when the frozen sea forms, according to Kauffman, is the phase transition from chaos to order.

What does all this self-organized complexity have to do with real life? Here is where Kauffman makes a daring leap. The central problem in the development of complex organisms such as ourselves is how one gets from an undifferentiated egg, by a process of repeated cell division, to a complete set of differentiated cell types—red blood cells, fibroblasts, retinal cones, and so on. Genes, Kauffman proposes, are units in a Boolean network. Cell types are attractors. And the cell

```
  8   8   1   1228228228228228228   1   1   1   1   1   1   1   1   1   1   1   1   1
  8   8   1   1   1   1   1228228228228   1   1   1   1   1   1   1   1   1   1   1   1
  8   8   845645645622822822822822822   1   1   1   1   1  10  10  10   1   1   1   1
  1   8   1   1228228228228   1   1   1   1   1   1   1   1  10  10  10   1   1   1   1
  1   1   1228228228228228228   1   1   1   1   1   1   1   1   1   1   1   1   1   1
  1   1   1   1   1228228228228228228   1   1   1   1   1   1   1   1   1   1   4   4
  1   1   1   1   1   1   1   1228228228228   1   1   1   1   1   1   1   1   1   1   1
  1   1   1   1   6   1   1228228228228   1   1   1   4   1   1   1   1   1   1   1   1
  1   4   1   6   6   6   1   1228228228228228228   4   1   4   1   1   1   1   1   1   1
  1   4   1   1   6   6   622822822822 8   1   1   1   4   1   4   1   1   1   1   1   1
  4   4   1   6   6   6   6   6228228228   1   1   1   1   1   1   1   1   1   1   1   4   4
  1   4  12   6   6   6   1228228228228   1   1   1   1   1   1   8   8   8   1   1   1   4
220   1   1   1   1   1   1   1   1228228228   1   1   1   1   1   8   8   8   8   1   1220
220220   1   1   1   1   1   1   1228228228228   1   1   1   1   8   8   4   8   1   1   1
220220   1   1   1   1   1   1   1228228   1   1   1   1   1   1   1   1   1   1220110   1
  1220110110   1   1   1   1   1228228   1   1   1   1   1   1   1   1   1  20  20110110
  1110110110   1   1   4   1   1228   1   1   2   4   1   1   1   1   1   1  20  20110110
110110110110110   1   4   1   1   1   1   1   2   4   1   1   1   1  20  20  20  20   1110
110110110  22   1   1   1   1   1   1   4   4228   1   1   1  20  20  20  20  20  20  20110
110110   1   1   1   1   1   1   1   1   1   1228   1   4   1  20  20  20  20  20  20  20110
110  22  22  22  22   1   1228228   1   1228228   1   4   4   1   1   1   1   4  20   2  22
 22  88  22  22   1   1   1   1228   1228228228   1   1   1   1   1   1   1   1  20   2   1
  1  88   1   1   1228228228228228228228228   1   1   1   1   1   1   1   4   4   4   1
  1   8   1   1228228228228228228228228228   1   1   1   1   1   1   1   1   1   1   1
```

6.2 Boolean net showing frozen component (indicated by 1's) and "twinkling" component (indicated by higher numbers—each number indicates the number of states in the cycle of that unit) (courtesy of Stuart Kauffman).

cycle is a state cycle. This means that during the interval from one cell division to the next, each particular cell type moves through a characteristic sequence of patterns of gene expression, each pattern leading by simple rules to the next one and returning to the starting point at the end of the cycle. Thus the complex specialization of cell types within a single animal or plant emerges "for free" from a few simple connectivity rules, with no need for an explicit description of what each cell type should be like. In fact, you could start by activating genes at random and you still would end up with the set of cell types characteristic for that genome, that is, for that species.

Do the numbers support this interpretation? According to Kauffman, they do. Genes generally are turned "on" or "off" by the activity of only a few other genes, perhaps two to five. And often these connections are canalizing functions. (An example would be a gene whose product suppresses the activity of a second gene no matter what state other genes may be in: a NOT IF Boolean gate.) So the connectivity among genes seems appropriate for producing orderly state cycles. The total number of genes, in humans, is thought to be around a hundred thousand. The number of cell types in the human body, Kauffman says, is 256, which is strikingly close to 316, the square root of 100,000. Experiments suggest that it takes about 1 to 10 minutes to turn a gene on or off, so it should take about 316 to 3160 minutes for a human cell to go once around its state cycle. In fact, dividing human cells do spend about this long between divisions.

As further numerical evidence in support of his hypothesis, Kauffman has presented a graph showing the relationship between the number of cell types and the total amount of DNA in an organism's genome, across organisms from bacteria and algae to mammals. On a logarithmic plot, the data follow a straight line, which Kauffman interprets to mean that the number of cell types in an organism does vary with the square root of the number of genes.

Finally, Kauffman cites studies suggesting that about 70 percent of the genes in the mammalian genome are permanently turned on in all cell types. These genes could represent the "frozen sea of lightbulbs" in an orderly Boolean network.

Still, it's pretty easy to pick holes in Kauffman's numbers. The number 256, for example, as an estimate of the number of cell types in the human body, Kauffman drew from a now-out-of-date cell-biology text. Yet if a "cell type" is defined as a cell that exhibits a unique pattern of gene expression, there are surely well over 256 types in our bodies. Just in the retina, for example, there's evidence for dozens of cell types by that criterion. "I agree with that worry," says Kauffman between mouthfuls of Greek salad. "I've agreed with it since I published this argument 30 years ago."

In addition, plotting the number of cell types against "total DNA" may be misleading, because as one goes to more-and-more-complex organisms, their genomes become encumbered with more and more "junk DNA" that has no known function. It's not even transcribed into RNA. The actual number of genes increases much more slowly than the total amount of DNA.

"Yes," says Kauffman, "If you look at the number of genes instead, the number of cell types goes up roughly linearly with the number of genes, rather than with the square root. But it's easy to devise a system where the scaling law is vastly different. It may be that in the long run all that will survive is the idea that you end up with a scaling relationship between the number of genes and the number of cell types. Then we'll find a category of genetic networks that obeys that scaling law, and that will be a reason to think it's a good category of networks."

The idea that 70 percent of all genes are turned "on" in all cell types is also open to question. "That really was the view 15 years ago," says Kauffman, "So that got written up in my books. We can find out much more precisely now, of course, and it may be less. I'm going to stick my neck out and say that the only way to be in the ordered regime is to have this frozen structure big enough to isolate the twinkling islands from each other, and therefore I bet that's true. But none of us mere theorists have looked hard at the mere experimentalist's naïve question: Why should this general principle actually be right? Evolution has had 4 billion years of doing very special things, so just because Stu and his friends come along and say that arbitrary networks behave like this, we don't know that a real network that's evolved over that period of time looks like that—we haven't taken one apart."

So what does actually happen in evolution? Traditionally, answering this question has been the business of paleontologists—people who study fossils. Before meeting with Kauffman, we get the chance to talk by telephone with two leading paleontologists, Stephen Jay Gould of Harvard and Simon Conway Morris of Cambridge University in England.

Gould and Conway Morris have both written books about the Burgess shale, the quarry in British Columbia where fossilized remains of the extraordinary creatures that lived in the Cambrian period were first unearthed. The two men share an intense fascination with the little soft monsters with their curious names—*Wiwaxia, Anomalocaris, Hallucigenia, Pikaia*—that crawled, swam, or burrowed their way across the floor of a shallow sea half a billion years ago (see Color Plate 13). Both men draw conclusions from the Cambrian fossils about the nature of evolution as a whole, but aside from their basic agreement that Darwinian evolution happens—which puts them at odds only with the creationists—their conclusions could hardly be more different. Perhaps most deeply, they differ in the emphasis they place on contingency—the role played by the chancelike concatenation of events—in guiding evolution's course.

Conway Morris played a key role in studying and classifying the Burgess creatures, as well as in finding other sites where Cambrian organisms are well preserved. In his book *Wonderful Life*, published in 1989, Gould lauded the much-younger British scientist to the sky, attributing to him the major discoveries and insights that allowed Gould to draw his major conclusion—that contingency is king and that evolution is therefore little more than a random walk through "morphospace." Conway Morris was brilliant, deserved a Nobel prize, and so forth. Since Gould is the Carl Sagan of biology, the great communicator whose books are instant best-sellers, his praise no doubt materially assisted Conway Morris's career.

Yet in his own 1998 book *The Crucible of Creation*, Conway Morris was unabashedly critical of Gould's arguments: they were "deeply flawed" and were characterized by "rhetoric," "invective," and an "ideological agenda." Contingency is not king, says Conway Morris; rather, there are forces in evolution that lead again and again to traits that we and other organisms possess. Life gravitates to some tiny corner of potential morphospace, because that's where the good designs are.

Even over the telephone, the two men are a study in contrasts. Gould offers assertive opinions about a wealth of topics and is inclined to peremptorily dismiss people or ideas, often with a characteristic snickering laugh. He says that thinking about extraterrestrial life is "fatuous speculation—it's one of the most useless, tendentious literatures ever." When asked how it happened that Conway Morris's viewpoint differs from the one depicted in *Wonderful Life*, Gould suggests that Conway Morris did have those views initially, but then retreated into orthodoxy. "You have to understand that Simon, when he did that early work, was a hippie graduate student, and now he's a very Establishment Oxbridge don."

Conway Morris, on the other hand, is all qualifications, hesitations, and confessions of his inability to make useful statements. "I'm not happy talking about things which I have no competence to explain," he says. "Why not?" we ask. "It's an old English failing," he replies. He is forever pointing out how much hand waving is involved in any particular line of thought. "It's a gin-and-tonic argument," he'll say, picking a very Establishment Oxbridge metaphor. About the only time that we get an unqualified opinion out of him is when the conversation wanders into a discussion of theories of multiple universes (see Chapter 10). "That" he says, "is total tosh."

The dispute between Gould and Conway Morris is one of those spats that enliven otherwise dull academic discourse. But it is also much more than that: It is deeply relevant to the theme of this book, for it goes to the heart of the question. Given that life evolves by Darwinian selection, what more can one say about it? Does evolution have any large-scale structures or principles beyond the "survival of the fittest"?

To Gould, the broad scientific and public acceptance of Darwinism came at the price of a distortion of Darwin's message.[6] As people gradually acknowledged that we were not divinely created on the Sixth Day, they reinvented their sense of specialness in a Darwinian mold. In this worldview, evolution is simply a way of getting *Homo sapiens* that doesn't involve miracles. The key principle of evolution is progress or upward striving. We start with boring bacteria, and things get steadily more complex, interesting, and diverse, until we finally and inevitably reach Ourselves. Gould points to traditional graphical depictions of the evolution of life on Earth, as in museum dioramas, which show worms replacing bacteria, fish replacing worms, frogs replacing fish, and so on. The implication of such depictions is that natural selection, by its incessant action at the level of individual organisms, steadily "improves" living organisms over the eons. Gould also points to scientific depictions of the "tree of life," or parts of that tree, as inverted cones, in which the variety of organisms' body plans increases steadily as time progresses.

Yet all this is wrong, says Gould. Life started simple, of course. So there had to be an initial drift toward increasing complexity: there was no other direction to go. But complexification is not a central theme of evolution. Bacteria have not gone away: they are still the most successful organisms on Earth. Fish are still far more successful and diverse than land vertebrates. Life is not getting steadily more complex. Some organisms, such as parasites, have simplified markedly over time. And life on Earth is actually *less* diverse now, in terms of the range of basic body plans, than it has been at some periods earlier in its history.

In 1972, Gould, with Niles Eldredge of the American Museum of Natural History, proposed the theory of "punctuated equilibrium."[7] This theory says that evolution has proceeded, not gradually and smoothly, but by fits and starts. Long periods of relative stasis were interrupted by short bursts of rapid extinction and speciation (appearance of new species). Such bursts might occur at various size scales, but recently attention has focused on the huge extinctions that ended several geological periods, most particularly the extinction at the boundary between the Cretaceous and Tertiary periods (the "K-T event") that wiped out the dinosaurs and a large fraction of other species. It's now widely agreed that this extinction event was caused by an asteroid impact.

The rules that governed which species survived the impact and which died out, says Gould, had little or nothing to do with survival during "regular" times. The rules changed suddenly and unpredictably. In no sense did the K-T event simply accelerate an ongoing process: it sent evolution on an entirely new path. So if the asteroid had missed the Earth by a few miles, the dinosaurs would still be in charge and *Homo sapiens* would be nonexistent and unimaginable. That asteroid impact was contingency operating at full throttle, but the same principle operates, Gould says, in small ways as well as large, just as it does in human history.

Contingency is not chance exactly, because determinism still rules supreme: it was the laws of Newtonian mechanics, after all, that brought the Earth and that asteroid to their fateful meeting. But the nature of evolution, according to Gould, is to magnify the effects of coincidental events, making the overall sequence of evolution (or human history) essentially unpredictable or chaotic. In Gould's metaphor, if one could replay the tape of life it would go forward quite differently from the way it actually did. And that means that any thought of humans as the expected culmination of evolution should be put out with the trash.

Why does the Burgess shale play such a crucial and contentious role in all of this? In *Wonderful Life*, Gould explains how Charles Walcott, who discovered the Burgess fauna in 1909, "shoehorned" the bizarre creatures into existing "phyla." (Phyla are the groupings of organisms immediately below "kingdoms;" they are sets of organisms with shared anatomical ground plans, such as sponges, arthropods, and chordates.) This, according to Gould, was because Walcott held to the notion of evolution as a goal-directed, upward expansion. Walcott could not see any point in phyla that went extinct, and so he refused to believe in them.

In the 1970s, Harry Whittington of Cambridge University, along with his graduate students Derek Briggs and Simon Conway Morris, began to reexamine the Burgess organisms.[8] As Gould tells it, their work led to the conclusion that many of the Burgess organisms were so different in morphology from any other known organisms that they must be assigned to entirely new phyla. The total diversity in basic body plans shown by the creatures in that one little Canadian quarry exceeded anything the world has seen before or since.

The Cambrian explosion—the first appearance of diverse multicellular organisms with hard parts in the fossil record—took place about 545 million years ago, at the beginning of the Cambrian geological period. The Burgess fauna date to about 520 million years ago. In

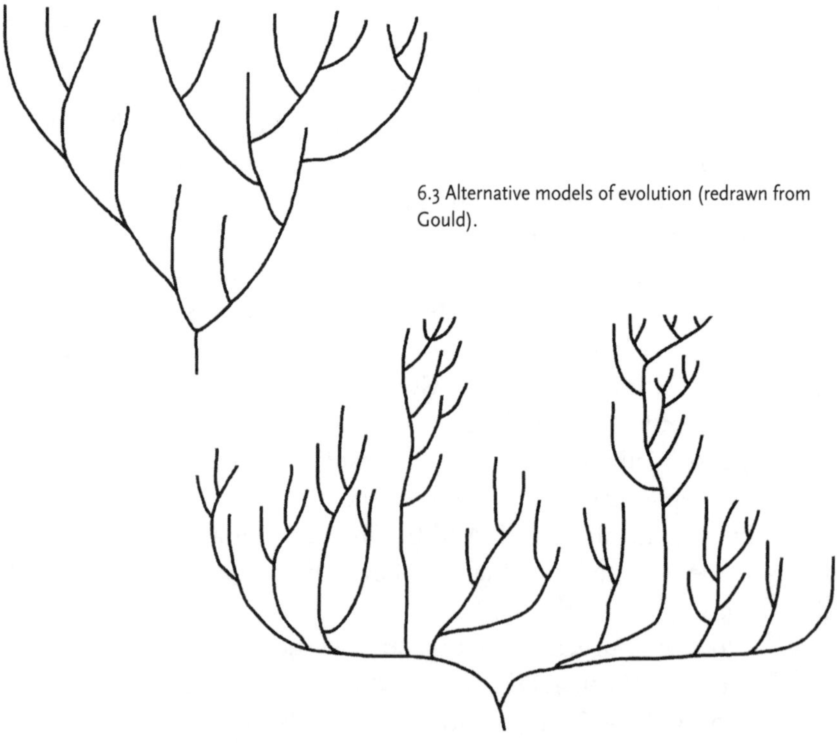

6.3 Alternative models of evolution (redrawn from Gould).

Gould's interpretation, the initial explosion represented a "filling of the empty barrel" with a wide variety of organisms. At the time of the Burgess fauna, then, life had reached its maximum diversity in terms of basic body types. Later, many body types were eliminated by natural selection. But, Gould argues, there's nothing to distinguish those body plans that "made it" from those that were subsequently eliminated. A scientist who had been present at that time simply could not have distinguished "better" from "worse" designs. The process of selection, though operating by Darwinian rules at an individual level, was essentially a lottery at a larger scale. Subsequent expansions, such as the one that took place after the K-T event, were quite different. The basic body plans already existed, and speciation therefore took place within existing phyla.

We ask Gould why he is so confident that the elimination of many Burgess types was a lottery. After all, we may not be able to see the good and bad points of the various body plans, especially as we have little notion of the conditions that the organisms faced. "Obviously, tomorrow someone might come up with a good reason why those that made it, made it," he concedes. "I think the main thing you can say in favor of the lottery is that there really is nothing that seems to unite the survivors. They don't share environment, they don't share similar structure, they don't even share a common location on the evolutionary tree. Suppose you found the survivors all in one part of the tree, then even though you might not be able to specify what it was, you could say, 'Hey, they're each others' relatives, they must share something.' But that's not true, they're scattered. It's not a proof."

Another possible criticism of Gould's thesis is that he's tilting at a long-derelict windmill. Who, after all, actually believes that a rerun of the tape would lead to humans? That idea has certainly existed in the past—it's exemplified in the novel *Auf Zwei Planeten,* mentioned in Chapter 3, in which Martians and humans, though the products of separate evolutionary lineages, end up so similar as to be capable of interbreeding. And maybe it's a notion that still has some credibility outside the biological community. But are there really biologists today who have that opinion? When we ask Gould to identify some, he comes up with one name: Simon Conway Morris.

"Gould is being a little bit mischievous, perhaps, in suggesting I think that," retorts Conway Morris. "As I try and explain in my book, *Homo sapiens* is a twig at the end of a long contingent pathway. But my argument would be that the name *Homo sapiens* in this sense is irrelevant. My suspicion is that the likelihood of some kind of intelligent species arising is probably very high. Whether they're humans or hydrocarbon arachnoids (to borrow a term from the philosopher Stephen Clark) really doesn't matter, it's the property of intelligence or moral responsibility, or whatever particular trait you're interested in, which matters."

Conway Morris presents arguments on several fronts. With respect to the Burgess fauna, he disputes the notion that there was greater diversity in basic body plans then than what is seen on Earth today. Part of the problem, he says, has to do with getting hung up on categories like "phyla," which, he says, represent "statements of taxonomic ignorance." If you choose your definition of a phylum in one way, you can argue that there were more phyla in the Cambrian period than there are today. But that doesn't mean that there was really more diversity of basic anatomy then than now. "My hunch is that, if anything, it's increased somewhat from then to now. Frankly, there were an awful lot of arthropods in the Cambrian, and they didn't have the range [of anatomies] that you see even in the crustaceans [just one of the many classes of arthropods] today."

To some extent, Walcott's "shoehorn" has made a comeback: some of the exotic Burgess creatures have been fitted, or tentatively assigned, to familiar groups. The most striking example is *Hallucigenia*. This was originally conceived as a tubelike animal with long spines for legs, on which it moved arthritically across the seafloor. It sported a row of short, enigmatic tentacles on its back. No remotely similar creature has existed before or since, and Gould argued that it probably constituted a separate phylum. More recently, however, better-preserved specimens of *Hallucigenia* and related species have been unearthed at other sites, which show that the beast actually had two rows of "tentacles" rather than one. They are in fact legs, and the spines stick upward from the back.[9] Having made this reversal, biologists immediately recognized *Hallucigenia* as a relative of an extant group, the onychophores, or "velvet worms."

"Simon Conway Morris described it upside down," Gould tells us, thus clarifying his own status as the victim of incorrect information. But *Hallucigenia* isn't the only Burgess creature that has crept back toward the fold of the familiar. Two other organisms that Gould claimed deserved novel phylum status are *Anomalocaris*, a large predator with massive grasping appendages and a wheellike jaw, and *Opabinia*, a smaller, vaguely crayfishlike beast with five eyes and a flexible, siphonlike proboscis. Citing work by paleontologists Graham Budd, Jan Bergström, and others, Conway Morris asserts that both of these animals are primitive arthropods, or close relatives of the creatures that gave rise to arthropods. *Opabinia* likely had arthropodlike legs, he says, but the legs are not well preserved.

Conway Morris thinks that convergence is an important feature of evolution. Not convergence of entire organisms, but of individual traits. The jaw of the marsupial saber-tooth "cat" of South America has an uncanny resemblance to that of the saber-tooth tiger of Eurasia. The lifestyles of marsupial and placental moles are strikingly similar. There are simply good ways of doing things, and so these ways crop up again and again, providing structure to evolution above the level of the mere "survival of the fittest."

Gould recognizes the existence of convergence, but plays down its extent and significance. The wings of bats and birds are not really convergent structures, he tells us, because they both arose from the vertebrate forelimb. The eyes of insects and vertebrates are not convergent structures, because there are developmental genes common to the eyes in both groups of animals, indicating that both kinds of eyes arose from the eye of some long-gone common ancestor. And so on.

Conway Morris also brings up developmental genes, but extracts a somewhat different message from them. He points to the discovery of the so-called *Hox* family of genes, which seem to play a basic role in organizing the body's basic layout.[10] *Hox* genes have been found in all animals from humans down to the humble cnidarians, simple two-layered organisms with stinging cells such as jelly-fish and sea anemones. Cnidarianlike fossils seem to be well represented in the Ediacaran fauna, a group of quilted, frondlike organisms that preceded the Cambrian explosion by several million years. But if the same family of genes has been used to produce the enormous diversity of animal body plans that we see in evolution, doesn't that mean that body plan may not be as fundamental an indicator of evolutionary change as paleontologists like to believe? At the very least, paleontology needs to be integrated with molecular biology. This is indeed a goal at which more and more biologists are aiming. Genomic data are obtainable only from extant organisms, however, and the attempt to work back to the genomes of, say, the Cambrian fauna is fraught with difficulty.

The topic of *Hox* genes actually illustrates rather well the possible differences between evolution at the level of observable traits and at the level of DNA. These genes have evolved by the accumulation of numerous single-base mutations, over the eons, consistent with Darwinian gradualism. In addition, however, the *Hox* genes have made some giant leaps. On the way from invertebrates to vertebrates, a single set of *Hox* genes doubled twice: we have four sets of *Hox* genes, whereas most invertebrates have one. Duplication of genes, followed by slow differentiation of the "daughter" genes, seems to be a common method by which the complexity of the genome increases. Yet the dramatic events—the duplications—may not themselves be accompanied by sudden changes in body plan. Conversely, a tiny, single-base mutation in one of the daughter genes may cause a radical change in the animal's morphology. Genetic evolution and phenotypic evolution (the kind you can see) do not necessarily track each other closely or follow the same styles.

To some extent, the differing viewpoints of Gould and Conway Morris represent a difference of emphasis. Gould acknowledges that evolution is not completely random: it's constrained by what is possible. "*Of course,* you don't get flying elephants," he'll say. For his part, Conway Morris acknowledges the existence of contingency as a factor in evolution. So one has to wonder whether, to some extent, the differences in interpretation flow from philosophical or even political differences that have little to do with the science.

Conway Morris clearly thinks so. "If I understand *Wonderful Life* correctly," he says, "the reason he wrote it, other than his genuine interest in our work—and I think that's unfeigned—was to suggest that humans were such an unlikely end product of evolution that, first, there's no question of directionality or progress, and linked to that the notion that here we are on our own and therefore it's up to us to sort out our own mess. Which is something that for other reasons I think is utterly wrong."

Those "other reasons" presumably include Conway Morris's religious faith. He believes in a God who didn't merely "turn the switches" at the time of the Big Bang, but who "permeates everything in a continuously creative way." He also believes in the power of prayer and in an after life. Thus, the quality of moral responsibility, which happens to be instantiated in the body of *Homo sapiens,* is a central part of God's design and is in no way a historical accident.

Aside from these fairly profound differences, there may be more quirky reasons for the two men's opposing viewpoints. Gould's interest in sports, especially in baseball, is legendary. The themes of winning and losing permeate his thought. Human history, he sometimes seems to believe, is a matter of who won and who lost, and why. That's what people are interested in, he tells us. He'll talk forever about how the battle of Gettysburg could have gone the other way and how everything would then have been different. His favorite metaphor is the horseshoe nail, for lack of which the kingdom was lost.

Certainly, horseshoe nails and asteroids are the final arbiters of history's course. But there is so much more. Kingdoms, for example—those amazing organizations that, utilizing vast energy resources in the form of taxation, deploy thousands of horses, each with four legs, and a shoe with six nails on each hoof, all to prevent that one missing nail from doing what it ultimately must do. Surely, there are questions about kingdoms, and the succession of kingdoms and other human institutions, that go beyond "What if King Harold had won the Battle of Hastings?" Perhaps Conway Morris is guided by an alternative principle, and one that's drummed into every British schoolboy: It's not whether you win or lose, it's how you play the game.

"I'm inclined to side with Gould," Stuart Kauffman tells us. To some extent, that's what we expected. Kauffman has built significant theoretical arguments on Gould's ideas. Punctuated equilibrium, for example, resonates well with complexity theory. Extinctions, large and small, are like the avalanches in Per Bak's computerized sand piles: they're a sign, Kauffman suggests, that evolution drives life toward a state of self-organized criticality, at the boundary of order and chaos. This offers an additional insight: mass extinctions don't have to have causes in the sense that we usually demand them. Yes, the K-T extinction was probably caused by an asteroid impact. But other extinction events may have been caused by the evolutionary equivalent of adding a single sand grain to the pile. In other words, the events were products of the evolutionary system itself.

Kauffman has also written a lot about the Cambrian explosion. He likes Gould's assertion that there was greater diversity of body plans in the Cambrian than there is today. In fact, in his book *At Home in the Universe*, Kauffmann estimates that three times as many phyla existed in the Cambrian as exist today. And he accepts that the Cambrian explosion was different from the expansions that followed the Permian and later extinctions. In the Cambrian, diversification was by formation of new fundamental body plans—phyla. But the later expansions involved diversification only at the tips of the tree—plenty of new species and families, but no new phyla or classes.

Kauffman finds a similarity between this evolutionary sequence and the typical process of technological innovation in human culture. Bicycles, automobiles, and airplanes, he says, began their careers by diversifying into all kinds of weird and wondrous variants, but later it came down to a matter of tinkering with the details. How true this really is perhaps deserves more detailed study than Kauffman has given it. After all, there is a tendency to mentally telescope lengthy, long-ago periods (when bicycles were first evolving, e.g.) into shorter spans of time. And nomenclature can rigidify before the product itself does, concealing the true extent of later innovation: an airplane whose wings revolve is not called an airplane but a helicopter, for example. But if we accept that there is a parallel between the processes of evolutionary and technological innovation, we naturally want to find explanations or rules that account for the similarity.

To study evolution, Kauffman uses "fitness landscapes." These are multidimensional arrays in which each location represents a possible organism. The multidimensionality represents the many ways in which an organism can vary—either at the level of its genes or at the level of the traits that the genes influence. The height of the landscape at each point represents the "fitness" of that possible organism: roughly speaking, how many offspring it is likely to leave. Kauffman studies how organisms, under the influence of mutation and selection, move across the landscape. In particular, how readily do organisms find the highest "peaks" in the landscape, the collection of traits that is optimal in that particular world? In their desire to move upward, how do organisms avoid getting trapped on local hilltops, far below the altitude of distant mountains?

These questions rephrase a familiar challenge to Darwin's "gradualism"—to his notion that you can get from here to there by tiny changes, each a slight improvement on the previous one and therefore favored by selection. The challenge, put forward by generations of creationists, is: How do you get complex structures like eyes? Wouldn't anything halfway toward an eye be useless and a hindrance, and wouldn't it therefore be selected against? Isn't fortress Eye surrounded by a deep defensive moat of functionless garbage?

Fitness landscapes have been a popular modeling tool for many years. In the 1970s, for example, Manfred Eigen of the University of Göttingen in West Germany used the landscapes to study the evolution of DNA sequences.[11] Eigen and his colleagues made a number of interesting observations. They showed that if the sequence is constantly mutating at some rate, it exists as a "cloud" of sequences at nearby locations on the landscape. The outlying members of this cloud are like scouts: they sense variations in the landscape at some distance from the region occupied by the majority. If the scouts find a higher location, they can pull the entire population in that direction, even across a small "valley" of lesser fitness. But this kind of behavior comes at a price: it can cause an "error catastrophe," in which the entire population drifts off its local peak and ends up worse off than where it began.

Kauffman's contribution has been to explore the ways in which tuning various parameters of a fitness landscape affect the ability of organisms to find the high peaks. In his "NK" models, the dimensions are reduced to binary alternatives. (An organism's "size," e.g., can be "small" or "large" but nothing in between.) The variable K has to do with an important property of organisms, which is that the value of any given trait is affected by other traits. Thus, the trait "winged/wingless" is influenced by the trait "small/large" (no flying elephants, remember). K is the average number of traits, out of the total traits N, that influence the fitness contribution of a given trait.

If K is set at zero (which never holds in real life), Kauffman gets a "Fujiyama" landscape, in which the height of each location is extremely similar to its neighbors—it's "highly correlated." By sensing the local gradient, an organism can head straight for the single, towering peak and find it without any hold up. If K is set very high, on the other hand, so that the value of a trait is influenced by most or all of the other traits, Kauffman gets uncorrelated "moonscapes," in which altitude varies randomly. On such a landscape, organisms quickly become trapped on local peaks and never approach the global optimum.

There is an intermediate domain of "rugged" landscapes in which Kauffman finds an interesting property, namely that the highest peaks can be reached from the largest number of possible starting points. This is obviously a favorable design feature in general, but it specifically helps answer the question: What makes sex so popular? Sexual reproduction, by generating mixtures of the genes of two unrelated individuals (recombination), produces offspring scattered across the fitness landscape between the locations of the two parents. Such widespread scattering can help the offspring find the highest peak, but only if random locations have a decent chance of lying on the lower slopes of that peak. With K tuned optimally, that condition is fulfilled.

What Kauffman is saying, then, is that evolving organisms may actively tune the parameters of the fitness landscapes on which they evolve. In a sense, the mechanism of evolution itself evolves. In evolving or not evolving wings, for example, an organism is not merely determining whether it will be able to fly, it is also tuning the density of trait-trait interactions (between wingedness and body size, e.g.) that ultimately control how it evolves in the future.

In real evolution, fitness landscapes change, not only because of genetic changes within each species, but because species influence each other in changing ways as they evolve. One animal develops strong jaws, so another develops a thick shell. That's coevolution. Kauffman has studied this process by connecting a number of NK landscapes, representing individual organisms, into a miniature ecosystem. In these experiments, Kauffman (with Kai Neumann) allows traits in one organism to influence the fitness contribution not only of other traits in the same organism, but also of traits in a variable number of "surrounding" organisms. By adjusting the various parameters of the model, Kauffman and Neumann can cause the ecosystem to settle rapidly into a stable state, or they can cause the organisms to evolve endlessly and chaotically as the organisms keep trying to adjust to changes in their neighbors. Remarkably, the model organisms tend to coevolve to a state, at the border of order and chaos, where the K values of individual organisms ensure the maximum average fitness of all the organisms in the ecosystem. This in turn minimizes the number of species that are driven to extinction. Another example, it seems, of how the mechanisms of evolution themselves evolve.

Powerful though Darwinian selection may be, Kauffman doubts that it can lead to the evolution of any arbitrarily chosen complex system. It may be, he suggests, that evolution has to make use of preexisting complexities of a kind that readily self-organize. If this is so, organisms will tend to return again and again to the robust patterns that reflect this "natural order," and evolution is more than just contingency. In reaching this conclusion, Kauffman may be more in sympathy with Conway Morris than with Gould.

As the bill arrives, Kauffman offers a final thought. "Where did these landscapes come from? They're not given by God. There's something much richer going on. Organisms play natural games; and as organisms evolved, so did the games they played. So the winning games are, of course, the games the winners play. Those are the games that can be explored well; they're ways of making a living that can be exploited by the search techniques you've got, which are mutation and recombination. So there's a self-consistent construction of a biosphere: autonomous agents making livings, ways of making livings, search procedures, creating niches for one another, and blossoming into existence. I think we have just the faintest images of what this is all about."

Of course, we ask both Gould and Conway Morris their opinion of Kauffman's work. We expect a very positive response from Gould, given that he supplied a glowing testimonial for the dust jacket of *At Home in the Universe*. In fact, however, he tells us that he likes Kauffman's work mainly as it relates to the origin of life. "Once you get into complex multicellular things, where you have a big influence of historical accidents, I'm not sure it's controlled the way Kauffman would have it." Conway Morris claims to be a "computer idiot" and adds: "So far I'm not persuaded that we have learned anything remarkable from the Santa Fe group, which is probably not a fair thing to say as I don't read their papers with monotonous regularity. What I don't sense is a simple connection to any sort of biological reality with which I am familiar."

How much chance? How much necessity? The debate goes on at every level of biology, from the fundamental details of life's chemistry to the traits that make us human. A good example at the chemical level concerns the genetic code, which governs translation between the four-letter language of RNA bases (U,C,A, and G) and the twenty-letter language of the amino acids that build proteins. Three-base triplets ("codons") represent amino acids: CGU codes for arginine, AAG codes for lysine, and so on. A completely arbitrary code, one would think, chosen by contingency and then "frozen" for all time on account of its inability to change gracefully without causing chaos.

In fact, however, molecular biologists started pointing out nonarbitrary aspects of the code as soon as it was discovered. For one thing, the third base in the codon does less than its fair share of work. If we look just at the first two bases in the codon, there are sixteen different possibilities (UU, UC, and so on), not far short of being enough to code for all 20 amino acids even without a third base. In fact, eight of these sixteen possibilities do specify a single amino acid, regardless of which base is the third member. UC specifies serine, for example, whether the complete codon is UCU, UCC, UCA, or UCG. And of the remaining eight 2-base combinations, several specify two amino acids that are functionally similar. Take the combination GA: GAU and GAC code for aspartic acid, and GAA and GAG code for glutamic acid. Aspartic acid and glutamic acid are so similar that they can often substitute for each other with no effect on the resulting protein's properties. Surely this kind of order is telling us something. For example, it seems likely that the code originally used doublets rather than triplets, with the third base serving merely as punctuation.[12]

More recently, the evolutionary biologists Laurence Hurst of the University of Bath, England and David Haig of Harvard, along with Hurst's graduate student Stephen Freeland, have performed computer analyses to investigate just how arbitrary (or not) the code may be.[13] They first generated a very large number of "random" genetic codes, and then calculated how robust each code was, that is, how often misreading a single base led to the incorporation of the same or a functionally similar amino acid. They found that they had to search a million random codes to find one as robust as the actual code. This they consider strong evidence that the actual code is not a "frozen accident" but has been molded by natural selection.

Perhaps even more strikingly, several researchers, including Laura Landweber and Rob Knight of Princeton University, have found that some amino acids have a chemical affinity with RNA sequences that

TTT:phenylalanine	TCT:serine	TAT:tyrosine	TGT:cysteine
TTC:phenylalanine	TCC:serine	TAC:tyrosine	TGC:cysteine
TTA:leucine	TCA:serine	TAA:[end]	TGA:[end]
TTG:leucine	TCG:serine	TAG:[end]	TGG:tryptophan
CTT:leucine	CCT:proline	CAT:histidine	CGT:arginine
CTC:leucine	CCC:proline	CAC:histidine	CGC:arginine
CTA:leucine	CCA:proline	CAA:glutamine	CGA:arginine
CTG:leucine	CCG:proline	CAG:glutamine	CGG:arginine
ATT:isoleucine	ACT:threonine	AAT:asparagine	AGT:serine
ATC:isoleucine	ACC:threonine	AAC:asparagine	AGC:serine
ATA:isoleucine	ACA:threonine	AAA:lysine	AGA:arginine
ATG:methionine	ACG:threonine	AAG:lysine	AGG:arginine
GTT:valine	GCT:alanine	GAT:aspartate	GGT:glycine
GTC:valine	GCC:alanine	GAC:aspartate	GGC:glycine
GTA:valine	GCA:alanine	GAA:glutamate	GGA:glycine
GTG:valine	GCG:alanine	GAG:glutamate	GGG:glycine

6.4 The DNA genetic code. In the RNA code, T (thymine) becomes U (uracil).

contain codons for those amino acids.[14] The amino acid arginine, for example, binds preferentially to RNA that is enriched in the six triplets that code for arginine (CGU, CGC, CGA, CGG, AGA, and AGG). This finding is quite a surprise, because the actual mechanism of translation doesn't involve binding between the codon and its amino acid; instead, it involves an interloper molecule called transfer RNA (tRNA). A possible interpretation of this result is that translation did originally involve codon and amino acid binding, and the tRNA interloper was added later to increase the specificity or efficiency of the process.

Again, we are led to the notion that the code is less arbitrary, and more influenced by chemical principles, than meets the eye. That's not to say that anyone seriously expects to find the identical genetic code beyond Earth, unless it be a Life that shares a common descent with our own. Even on Earth, a few microorganisms use a slightly different code, as do the mitochondria in eukaryotic cells. We don't even have any strong assurance that life elsewhere uses the DNA-RNA-protein cascade. But the possibilities may not be as endless as one might imagine.

To choose an example from the other end of the scale, what about intelligence—specifically, the "high intelligence" with which we credit *Homo sapiens*? If we leave aside the question of what kind of physical organism this trait may reside in, is the trait itself so intrinsically valuable that evolution is likely to discover it one way or another?

This is a case where the first-glance answer is likely to be yes. It may take a while—it took over 3.5 billion years on Earth—but eventually it will appear, and once it does appear in some degree, it will be strongly selected for ultimately offering its owners so powerful a benefit that they will make themselves lords of the Earth in perpetuity.

Not everyone shares that view, though. Ernst Mayr, the doyen of evolutionary studies at Harvard, has put forward an opposing idea: the tree of life, he says, is a many-budded bush in which organisms with high intelligence—namely, ourselves—occupy one insignificant twig.[15] One species out of billions of species, since the origin of life on Earth, has possessed this trait. So how could one possibly assert that evolution has in any sense favored its development? Stephen Gould makes similar arguments.

Another evolutionary biologist, Jared Diamond of UCLA, has pursued this same idea by drawing an analogy between the traits of high intelligence and woodpecking.[16] Woodpeckers occupy such an obviously profitable niche that many species, one would think, would have converged on it. But, in fact, all woodpeckers belong to a single lineage. The niche could easily have remained unfilled—and, in fact, does remain unfilled in parts of the world, such as remote islands, where woodpeckers can't fly to. It's the same with high intelligence, says Diamond. It looks inevitable with hindsight— but ahead of time, it's highly improbable.

Unquestionably, Mayr and Diamond help restore some balance to our warped sense of our own importance in evolution. On the other hand, there's something a bit artificial about separating the trait of "high" intelligence from mental capacities in general. If we look over the Earth, it's surely fair to see a great blossoming of complexity in the way organisms interact with their surroundings. This blossoming of behavioral and mental complexity is supported by a widespread elaboration of molecular pathways and neurological circuits. Bacteria can turn genes "on" or "off" in response to the availability of foodstuffs, sea slugs can be conditioned to "remember" painful stimuli, parrots can classify objects, chimpanzees can invent uses for tools, we can think about life in the cosmos. In other words, "high" intelligence may not be as isolated a trait as Mayr and Diamond seem to suggest. "If Gould is right, it may turn out that intelligence is an incredibly rare property," says Simon Conway Morris, "but the very fact that we have intelligence in mollusks, especially in octopus and squid, makes it seem that it's something rather probable. Even protozoa, they have incredibly sophisticated systems of chemical response, even in a single cell."

No trait's value can really be assessed in isolation, of course. Thus, while few, if any, scientists consider human beings to be the necessary vehicle for intelligence, there's a widespread view that certain traits, or behavior patterns that the predecessors of humans possessed, were important for the later development of intelligence. Neurophysiologist William Calvin of the University of Washington, for example, believes that our intelligence arose out of the demands of planning rapid or precise movements.[17] Such a view would seem to predict that intelligent organisms in general should have well-developed manipulative or throwing skills, as humans do. On this question, Conway Morris is his usual cautious self. "As soon as the flying saucer arrives and the bipeds come down, we'll say, 'I told you so,' " he comments. "And as soon as the flying saucer arrives and out comes a thing like a caterpillar, we'll say, 'I told you so.' There isn't any way of deciding."

* * * * *

What does all this say about the course of evolution in other parts of the cosmos? To Gould, it says very little, except that we shouldn't expect to find human beings or even intelligence. He refuses point-blank to speculate about life elsewhere, even as to quite basic questions such as whether extraterrestrial life should also evolve by the Darwinian mechanism, or whether one could expect to find sexual reproduction. "I'm not going to play," he says. "I know pop literature loves this, and you'll probably sell more books, but I don't think it's useful to speculate about whether the way it happened here is the only way it could happen."

Since Kauffman laces his writings with mantralike expressions such as "At Home in the Universe" and "We the Expected," he obviously believes in the existence of other Lifes (although the "we" must refer to complex life-forms in general, rather than to specifically anthropoid creatures). When we ask him about extraterrestrial life, on the way back to the Bios Group office, he says: "I think I estimated about one or two per galaxy." After we express some surprise at the low estimate, he adds: "Maybe that was the ones with intelligent civilizations." Evidently Kauffman, like Gould, doesn't consider it worth spending too much time trying to make detailed predictions about what we may find out there.

Mayr and Diamond believe that intelligent life is so rare in the universe as not to be worth looking for. One might as well go looking for woodpeckers, says Diamond. Yet the analogy may not be quite appropriate. Woodpecking, after all, requires the prior existence of two other specific kinds of organisms: (1) trees and (2) bugs that live under the bark of trees. Intelligence seems like a more general trait that could exist in many different kinds of biospheres.

Simon Conway Morris believes that, where life evolves, intelligence has a very good chance of evolving too. But because we haven't heard anything from extraterrestrials, he is inclined to think that intelligent creatures must be rare (this line of thought is taken up in more detail in the next chapter). Putting these two ideas together, Conway Morris surmises that life itself may be very rare indeed in the universe—otherwise we would have heard from it.

What all these scientists agree on, however, is the value of looking for life beyond Earth. Gould even supports the search for radio signals from alien civilizations, in spite of his doubts about the inevitability of intelligence. "Although the probability of success is very very low," he tells us, "the results are of such potential importance that I think it's worth spending a little money on, because it would at least be direct evidence."

Gould's famous question, What would happen if we replayed the tape of life on Earth? is to some degree an invitation to the very kind of speculation that Gould detests, for in fact we never can replay the tape. Beyond Earth, however, that tape may have been replayed times beyond counting. Beyond Earth, therefore, may be where his question will be answered.

As we wander back to our car, our heads are spinning. It's hard to know which of all the ideas we've heard will turn out to be enduring and important, and which ephemeral; harder, perhaps, than picking out the natural turquoise and lapis lazuli amongst all the plastic jewelry laid out in the Santa Fe plaza. But Kauffman has at least persuaded us that there are gems worth searching for—that there may indeed be laws of complexity that help chart the course of life beyond the immediacy of Darwinian selection. And Kauffman's own assessment, with its characteristic jumble of pride and modesty, continues to ring in our ears. "I do think it's important—most biologists still are ignoring what I've talked about, but I don't think that they will in the next couple of decades. I don't say that with complete confidence. I say it with a lot of humility, having been ignored for 30 years."

SUGGESTED READING

Bengtson, S., ed. *Early Life on Earth.* New York: Columbia University Press, 1994.

Conway Morris, S. *The Crucible of Creation: The Burgess Shale and the Rise of Animals.* Oxford: Oxford University Press, 1998.

Gould, S.J. *Wonderful Life: The Burgess Shale and the Nature of History.* New York: W.W. Norton, 1989.

Kauffman, S. *At Home in the Universe: The Search for the Laws of Self-Organization and Complexity.* New York: Oxford University Press, 1995.

7

SETI

The Search for Extraterrestrial Intelligence

By the year 2000?" Frank Drake laughs. "When I started saying that, the year 2000 seemed a long way off! I'm starting to weasel out of that now—but there's still a chance."

Making contact with extraterrestrials has been Drake's lifelong dream, and one he's not likely to abandon just because of the arrival of the millenium (and his own seventieth birthday). The genial, silver-haired radio astronomer has been sitting by the cosmic telephone since 1960, waiting for the call that will change everything forever. Deadlines have come and gone, funding has waxed and waned, researchers have entered and left the field, but Frank Drake has been a fixture—lone pioneer, tireless booster, and now respected elder of what sometimes seems more like a movement than a field of science: the search for extraterrestrial intelligence, or SETI.

We talk with Drake in his office at the SETI Institute, of which he is president. Located in an office park in Mountain View, California, conveniently close to Stanford University, Silicon Valley, and NASA Ames Research Center, the Institute resembles an upscale version of the institute for Creation Research in San Diego. The institute's lobby, like the Creation Museum, displays NASA photographs of the cosmos. But it also has something that the creationists can't boast of—a signed photograph of actress Jodie Foster, star of the SETI movie *Contact,* standing in front of the Arecibo radio telescope. "Thanks for all your help!" she has scrawled across it.

Hollywood, once obsessed with all things biblical, now locates its epics on the frontiers of science. Innumerable movies have dealt with intelligent extraterrestrials in one way or another. But *Contact,* based as it is on a novel by Carl Sagan, is unusual for its attempt at a scientifically realistic portrayal of the SETI enterprise. The SETI Institute's website devotes a page to the film, pointing out that some of the people in it, including Jodie Foster's character, are based on real-life researchers at the institute. It also corrects some of the film's minor scientific howlers: the idea, for example, that when computers are flummoxed, humans can decipher extraterrestrial signals simply by listening to them with headphones. If the institute has any doubts about the plausibility of the film's ending, however, in which Foster travels via cosmic wormholes to a galaxy far, far away, it keeps quiet about them.

A certain mutualism exists between SETI and the entertainment industry. The astronomers' most photogenic radio telescopes—Arecibo, the Very Large Array at Socorro, New Mexico, and the Owens Valley Radio Observatory—have been put at the directors' disposal and have become familiar screen icons. In return, Hollywood has supported SETI. Steven Spielberg, for example, has funded SETI at Harvard University. Arthur C. Clarke, author of *2001: A Space Odyssey* and many other highly regarded works of science fiction, supports SETI at Berkeley and elsewhere and helps raise private money for SETI projects.

The main support for the SETI Institute, however, comes from the electronics and computer industry. According to Drake, Microsoft's cofounder Paul Allen and Intel's president Gordon Moore each give the institute about a million dollars a year. The Hewlett-Packard Company has been a particular gold mine: Bill Hewlett and the late David Packard have both been major supporters. Hewlett-Packard's late Vice President for Research, Bernard Oliver, who was actively involved in designing SETI projects during a stint at NASA Ames, left about $20 million to the institute.

It is just as well that SETI has this degree of private support, because the federal government bowed out of the SETI business in 1993. On October 1 of that year, Senator Richard Bryan of Nevada rose in

Congress to denounce the fledgling NASA program. "As of today, millions have been spent, and we have yet to bag a single little green fellow," he said. SETI funding was promptly axed and has never been restored, but Drake remains convinced that SETI is broadly supported by the public and by the scientific community.

The idea of communicating with extraterrestrial beings goes back a long way.[1] In the nineteenth century, the German physicist Karl Friedrich Gauss suggested that a giant right-angled triangle be cut into the forests of Siberia, thus advertising our knowledge of geometry to alien astronomers. Radio pioneers Nikola Tesla and Guglielmo Marconi both believed that they had actually detected signals from other worlds. During Mars's 1924 opposition, American astronomer David Todd asked for two days' global radio silence, in order that wireless operators might listen for messages from that planet. During the silence (which was only partially honored), stations at Vancouver, British Columbia, and in London, England, both reported picking up signals transmitted in an undecipherable code. If Martians were indeed trying to contact us, however, they must later have changed their minds and gone into hiding.

The modern SETI era began in 1959, when two Cornell University physicists, Giuseppe Cocconi and Philip Morrison, published a short paper in *Nature* entitled "Searching for Interstellar Communications."[2] Civilizations much older than our own, they suggested, might be directing signals toward us. These signals would most likely be transmitted in the radio band of the electromagnetic spectrum. Although there was a dishearteningly wide range of frequencies to choose from within the radio band, one particular frequency, 1420 megahertz (corresponding to a wavelength of 21.1 centimeters), seemed like an obvious candidate for an interstellar beacon, because it would be universally recognized as the frequency of radiation from neutral hydrogen atoms in the interstellar medium. (A hydrogen atom emits a 21.1-cm photon when the "spins" of its proton and electron flip between the parallel and antiparallel states.) Cocconi and Morrison calculated that, even with current radio telescope technology, communication over interstellar distances was feasible, and they urged radio astronomers to begin the search.

During the very months when Cocconi and Morrison were preparing their manuscript, Frank Drake had independently come to similar conclusions. As a young astronomer at the National Radio Astronomy Observatory in Green Bank, West Virginia, Drake had the means to realize his plans. With the support of the observatory's director, Otto Struve, Drake began observations on April 8, 1960, by aiming the observatory's 85-foot dish at two nearby Sun-like stars, Tau Ceti and Epsilon Eridani. (As mentioned in Chapter 4, the latter star is now suspected of possessing at least one planet.) Using a single-channel amplifier hooked up to a chart recorder, Drake explored frequencies near 1420 megahertz; but aside from one false alarm, he detected nothing.

These early theoretical and practical efforts succeeded in drawing considerable attention to the subject; and in the following year, Drake and Struve organized a meeting of interested scientists to discuss the whole question of extraterrestrial intelligence and how best to search for it. The meeting was attended by such diverse figures as biochemist Melvin Calvin, dolphin-intelligence researcher John Lilly, physicists Cocconi and Morrison, astronomer Carl Sagan, and electrical engineer Bernard Oliver. At the meeting, Drake introduced what has since become famous as the "Drake Equation":

$$N = R^* f_p n_e f_l f_i f_c L$$

The equation says that N, the number of communicating civilizations within the Milky Way galaxy, is equal to R, the number of new stars formed each year, multiplied by a number of factors: f_p, the fraction of stars that have planets; n_e, the average number of habitable planets per star; f_l, the fraction of such planets on which life actually developed; f_i, the fraction of those planets on which intelligence arose; f_c, the fraction of those planets on which intelligent organisms communicate across space; and L, the average lifetime of such communicating civilizations.

Drake proposed that the various factors were either near 1 or canceled each other out, so that the equation reduced to $N = L$, that is, the number of communicating civilizations in the galaxy at any one time is roughly equal to the average lifetime of such civilizations. He suggested that a reasonable estimate of L was 10,000 years, so that one could expect there to be about 10,000 civilizations in our galaxy right now, all busy sending signals into space.

The Drake Equation has aroused mixed reviews over the years. To some enthusiasts, it is up there with $E = mc^2$ in the pantheon of world-altering insights. To others, it is the worst of pseudoscience. "It's not an equation," Stephen Gould tells us baldly, adding his trademark chuckle to let us know that he has finally put the thing out of its misery.

To us, the equation seems most useful simply as a framework for discussing the many conditions that need to be fulfilled before we can count on making contact with extraterrestrial civilizations. As Drake tells us, some of the conditions now appear more probable than they did in 1961: extrasolar planets have been detected, and the origin of life seems much less like a freak event than it did then. But the quantity L, the lifetime of an average civilization, still seems like anyone's guess.

Drake doesn't see it that way. "It's a little more sophisticated than people realize," he tells us. "If you do the mathematics, it turns out that the value of L is very strongly dominated by any very long-lived civilizations. If only 1 percent of them remain visible for a billion years, then L will be 10^7 years, even if all the other 99 percent blow themselves up overnight."

On Earth, of course, no species has survived for anything like a billion years, let alone maintained a technological civilization for that long. If anything, making L depend on the existence of a few extraordinarily long-lived civilizations seems to make the estimate of N less trustworthy than otherwise. But Drake exudes cheerful confidence. His car license plate reads: "N EQLS L." "It's a political statement," he tells us with a smile. "It's saying that the people who say we're alone are wrong."

The story of SETI since 1961 has been one of gradual technical and conceptual advances. Bernard Oliver, for example, stressed that the 1420-megahertz hydrogen line is not the only possible "beacon frequency"—he pointed out the lines at 1612, 1665, 1667 and 1721 megahertz, due to the hydroxyl (OH) radical, as alternative candidates. Since OH and H make water, Oliver dubbed the entire band between 1420 and 1721 the "water hole"—a quiet region in the spectrum where galactic civilizations might meet.[3] A number of other "magic frequencies" have been suggested. Even the notion that signals would be sent at radio frequencies has been questioned. By transmitting extremely brief pulses, for example, intelligent beings might be able to outshine their parent star even at optical wavelengths.

Another question that has aroused a lot of discussion is whether we should be looking for intentional transmissions—galactic beacons—or for radiation emitted as an unintended by-product of technologies such as television and radar. The latter sources might be relatively common, but they would also be relatively weak and would not be at "magic frequencies." What's more, there's no guarantee that civilizations do leak radiation into space for long periods, the way we are doing now. "The thing that is ringing alarm bells for me," says Drake, "is that we see our civilization going very rapidly toward the use of fiber-optic systems, and to direct-to-home satellites. A typical TV station radiates a million watts, but a typical satellite transmits at 100 watts, and of that only about 10 watts leak out into space. So we're rapidly losing visibility—by a factor of 100,000. Is that typical or quirky? We don't know, but it's a warning signal."

A major problem for SETI researchers has always been the availability of telescopes. Because there is so much sky to search, and so many frequencies to scan, a serious SETI search can easily consume thousands of hours of observing time per year. But large radio telescopes are in tremendous demand for conventional astronomical research. Even Drake, who was for many years director of the Arecibo telescope, has never been able to command the amount of telescope time that is really needed. So SETI enthusiasts have long made plans to build their own, special-purpose instruments. Bernard Oliver, for example, while he was at NASA Ames Research Center in the early 1970s, came up with a scheme called Project Cyclops. This was a plan to construct an array of up to a thousand linked radio telescopes, each 100 meters in diameter. The instrument would have been capable of picking up signals from anywhere in the galaxy. The multibillion-dollar project never became reality; in fact, its grandiosity may have helped sour the atmosphere for public support of SETI programs.

Given unlimited funds, Drake would not resurrect Project Cyclops today, mainly because of a change in the relative costs of radio telescopes and computers. "Radio telescopes have always cost the same as hamburger, per pound," he says. "But the weight of a radio telescope goes up with the cube of the radius, whereas the collecting area goes up only with the square of the radius. So building large telescopes costs more per square meter of collecting area. Now the cost of computer technology has fallen to the point that it's more economical to link large numbers of small telescopes. Right now, we're looking at a plan to buy three thousand small dishes, like the dishes people have in their back yards. They can be linked to make the equivalent of the Arecibo telescope. And Arecibo looks only within 20 degrees of straight up, whereas with these arrays you can look anywhere you want, and you can even have multiple beams, looking at different parts of the sky at the same time."

Before it was axed by Congress, NASA's SETI program had two arms. One arm—the "targeted search"—involved a detailed study of about a thousand Sun-like stars located within 100 light-years of Earth, using large telescopes such as Arecibo. Each star was to be studied for about 10 minutes, over a frequency range between 1000 and 3000 megahertz (i.e., bracketing the "water hole"). The other arm was a stepwise survey of the entire sky, without regard to the locations of particular stars, using smaller telescopes belonging to the federal government's Deep Space Network. A wider range of frequencies, from 1000 to 10,000 megahertz, would be studied, but the observation of any particular patch of sky would be quite brief, so that the sensitivity and frequency resolution would be low.

These two approaches represented opposite underlying philosophies. The targeted search was based on the premise that transmitting civilizations are quite common, so that even a sample of a thousand stars had some chance of including one of these civilizations. The all-sky survey was based on the idea that the galaxy contains a small number of powerful transmitters, none of which are likely to be located near Earth.

After the NASA program ceased, the all-sky survey became impracticable, because the government's Deep Space Network was now off limits. Thus, the SETI Institute has pursued the other arm, the targeted search. To signify its rebirth from the ashes of the NASA program, the search has been named Project Phoenix.[4] Drake takes us downstairs and introduces us to the project's director, Jill Tarter.

Tarter is a strikingly tall and handsome woman. Her gaze is attentive, almost severe; her diction is precise, her arguments clearly expressed. In short, she is the very opposite of a flaky "bagger of little green fellows." At the time of Senator Bryan's intervention, she was the project scientist for NASA's budding SETI program, so the termination of the program, and Bryan's remarks, were very traumatic—"like getting hit in the stomach," she tells us. Luckily, the SETI Institute has been a safe haven for her: she now occupies the "Oliver Chair," one of the two positions endowed from Bernard Oliver's bequest (the other, the "Sagan Chair," is held by Chris Chyba).

Tarter is one of the world's few full-time SETI researchers. Frank Drake, though long identified with SETI, always had a "day job" as a regular radio astronomer or observatory director. In fact, Drake has said on several occasions that SETI is not an appropriate full-time occupation for a scientist, because spending years searching for a signal, and not finding one, would not be sufficiently challenging or career advancing. Tarter has an even more sober assessment of SETI's prospects than does Drake. Whereas Drake has predicted success before the millenium, Tarter has said at least once that she doesn't expect success in her own lifetime. So what keeps her going?

We broach this subject by mentioning her earlier theoretical work on brown dwarfs. If she had stuck with them, we suggest, she would have experienced the reward of their discovery. "When I was working on brown dwarfs," she replies, "I was really happy doing what I was doing, but I sometimes wondered, 'What was the US taxpayer getting out of this?' When I began to work on SETI that whole question went away because now I was working on something that I knew the person in the street had an interest in. Not everybody is positive, but I knew it was a project that was meaningful. It could in fact make an enormous difference, so I've never looked back. Even if we don't succeed in our lifetime, I will have been part of something that establishes a search and allows it to go on for as long as it may need to. And do I really think it won't happen in my lifetime? If you look in that refrigerator over there, you'll see a bottle of champagne."

While the champagne continues chilling, we walk across the hall to the institute's nerve center, a nondescript room lined with terminals along one side. The computers are linked to the 43-meter telescope at Green Bank, West Virginia, as well as to a smaller secondary telescope in Woodbury, Georgia. The degree of automation is impressive. The computer selects a star to study from a catalog of Sun-like stars, taking into account the prior history of observations and the area of the sky that will be in view during the following few hours. The computer then points the Green Bank telescope at the star and tracks it as it moves westward across the sky. At any one time, the computer is examining a 20-megahertz-frequency band, which it divides into 56 million channels, each 0.7-hertz wide and at one of two orthogonal polarizations.

The width of the channel—in this case, 0.7 hertz—is an important variable. Generally speaking, natural signals are broad-band, while artificial signals are expected to be narrowband, so a basic winnowing procedure is to look for signals that are present only in narrow frequency channels. How narrow? If extraterrestrials are intentionally trying to contact us over galactic distances, they would do best to put all their power into as narrow a band as possible. Such a signal will be optimally detectable because it has to compete only with "noise"

7.1 Project Phoenix detects an extraterrestrial signal. The vertical axis of the computer display is time: new rows of data are continuously added at the top, so that the display descends like a waterfall. The horizontal axis is frequency, with low frequencies to the left. The dots represent frequencies where the radio noise exceeds some arbitrary threshold. The alignment of dots into a line, at lower left, reveals the signal from the Pioneer 10 spacecraft. The slant of the line shows that the frequency of the signal is decreasing over time: this is a Doppler effect caused by the Earth's rotation (courtesy of Jill Tarter).

within its own band. The more the power is distributed over many frequencies, the more noise it has to compete with. There is, however, a limit to the narrowness of the optimal channel because radio signals become slightly dispersed in frequency as they encounter electrons in interstellar space. The dispersion is of the order of 0.1 hertz, so sending or receiving in frequency channels narrower than about 0.1 hertz is pointless.

There is also a trade-off between detectability and information content. The width of a channel determines how much information can be carried by that channel because the information is contained in the modulation of the basic carrier wave, not in the carrier wave itself. Because of this trade-off, it's often been suggested that an intentional extraterrestrial signal would have two parts: a powerful, narrowband "beacon" signal and a weaker broadband "message" signal. The beacon signal might lie within the "message" channel, or it might contain information about where to find the message channel.

From the point of view of the search technology, there is another trade-off: the narrower the channels that are studied, the more channels that need to be observed if one wants complete frequency coverage. Hence SETI researchers have invested a great deal of effort into the development of multichannel analyzers.

When the SETI Institute's computer has acquired a couple of seconds' worth of signal from a set of 56 million channels, it displays the data (or a sampling thereof) on a horizontal line on the screen. Ascending frequencies are represented along the line. A dot appears at any frequency at which the signal strength exceeds some fairly low threshold. New rows of dots are continually being added at the top of the display, so the whole pattern of random-looking dots descends the screen continually like a waterfall, with frequency in the horizontal dimension and time in the vertical dimension. Most of the dots are caused by random variations in noise level, and thus resemble "snow" on the screen. A real signal, however, shows up as a streak of dots running at some angle across the screen. If the frequency of the signal is unchanging, the streak will be vertical. For most extraterrestrial sources (such as signals from orbiting satellites or space probes), the streak will not be vertical but oblique, indicating that the signal's frequency is drifting higher or lower over time. This drift is a Doppler effect, caused by gradual changes in the relative velocity of the source and the receiver. In particular, the Earth's rotation causes a predictable downward drift in frequency as a distant source traverses the sky, and this downward drift levels out as the source sets in the west, because the telescope is then moving at a constant speed directly away from the source.

While continuing to acquire new data, the computer searches the previously acquired batch of data for "streaks." When the computer finds a "streak," it compares it with a database of previously identified sources. If it matches a known source, the computer assumes that the signal is not coming from the region of sky at which the telescope is pointing, but from a "sidelobe." (Because radio telescopes collect radiation at wavelengths that are large compared to the dimensions of the reflector, they suffer from significant diffraction effects, such as picking up extraneous signals at certain large angles from the straight-ahead direction). If the signal is judged to come from a sidelobe, it is rejected.

If the signal is a novel one, on the other hand, the computer goes back to the frequency band at which the observation was made in order to see if the source is still there. If so, the computer commands the secondary telescope in Georgia to look for the source. If that telescope finds the same signal, the computer compares the Doppler drift of the signal as received by the two telescopes. Because the two telescopes are at different locations on the Earth's surface, the Doppler drift measured at the two telescopes should differ in a predictable fashion. If this condition is fulfilled, the computer proceeds to its next test, which is to repeatedly move the Green Bank telescope away from the star and back onto it. If the signal is recorded only while the telescope is pointed at the star, then it is finally "bells and whistles time," as Tarter puts it. "Not only bells and whistles here, but the computer sends messages to all our pagers and cell phones and sends E-mail to everybody."

During an earlier run, when Tarter and her colleagues were using a pair of telescopes in Australia, the bells and whistles went off thirty-nine times. These events, however, invariably occurred when the signals happened to be located along a line of ambiguity between the two telescopes, where the predicted Doppler drift was the same at the two instruments. In other words, the computer was unable in those cases to properly distinguish an extraterrestrial source from a side-lobe source. The computer was re-programmed to ignore signals along this line of ambiguity, and not a single bells-and-whistles event has taken place since that time.

We ask Tarter about how she would handle an apparently genuine event. Ideally, she says, the finding would be kept confidential while radio observatories around the world were given a chance to confirm the observation. A portion of the sky around the pole star is always visible, but most regions of the sky set after a few hours, so she would notify all the world's observatories, using the International Astronomical Union telegram system, in order that the source could be confirmed and monitored continuously. Once the source had been confirmed by other telescopes and clearly identified as artificial and extraterrestrial, it would be time to hold a press conference and tell the world. "The signal isn't being sent to California," Tarter says. "It's being sent to planet Earth, and everyone deserves to know."

Of course the best-laid plans may not work out in the heat of the event. Tarter recounts an episode that took place in June 1997. "It was on a day when the Woodbury telescope had taken a lightning strike, which took out the disk drives. Since we had already paid for Green Bank, we were using it in a mode where you move the telescope alternately between two stars, and one star serves as the off position for the other. We got a signal that we saw when we looked at one star, and we didn't see it when we looked at the other star. And we didn't see it when we moved the telescope off target, and when we went back on target, it was there again. And so it went. I had agreed with a documentary film crew that they could come and film that day. Well, about 10 minutes after we got into this, the film crew walk in the door, and there goes our confidentiality protocol out of the window. What do I say, 'You can't film'? For hours the source did the right thing, but by the time it set, the differential Doppler was looking wrong. We spent a lot of time on the web that night, and the next morning everybody was here, but we were able to detect the source even before it should have risen, and that's because it was the SOHO spacecraft, coming in through a side lobe 90 degrees away from where we were looking." No one was more disappointed, presumably, than the members of the film crew, who thought they had stumbled on the scoop to end all scoops.

One criticism that's been leveled against SETI is that its basic hypothesis is unfalsifiable: however many stars are examined, or how many frequencies tested, there will remain infinitely more to be sampled. "At some point society is just going to say, 'There's enough effort been put into this. For all practical purposes, we are alone,'" says Tarter. "But we do try to delineate the hypotheses. To date, we can say that we've looked at more than four hundred solar-type stars in our neighborhood; we've looked at frequencies from 1 to 3 gigaherz, and at the time we examined those stars we found no transmitters as strong or stronger than our large military radars. We could say that we don't yet have the sensitivity to detect television and FM leakage for all our targets—in fact, we can detect that just to the nearest star, barely. On the other hand, if you talk about the Earth's strongest signal, which is the planetary radar associated with the Arecibo observatory, our system could detect signals of that strength halfway to the center of the galaxy. We're doing what we can in a stepwise fashion; and as we learn more, we may change what we do, and we'll be able to do what we do, better."

We can't resist asking Tarter about Ellie Arroway, the central figure in *Contact*. "When I read the book," she says, "I thought, Well, this is awfully close to home. I certainly went through a lot of things that Ellie did as a young woman trying to do a nontraditional career, had the same sorts of problems as a young researcher with more-senior established people trying to take credit for things that were done. That's part of the 'who-gets-to-go' plot in the book and the movie. But that wasn't unique to me." Tarter tells us about a 1978 meeting of women scientists who had recently obtained their PhDs. "We did a lot of demographics, what it was that helped these women succeed—to ignore the messages they'd received growing up. Most important was a mentor early on, somebody who really said you could do it. Often it was the father, and not infrequently the father died. My own father died when I was 12 years old. I guess you're left with the 'I'm going to do this for my Dad' kind of stick-to-it-iveness. The other interesting thing we learned was that most of us had been either cheerleaders or drum majorettes in high school, which is so absurd, except that you think that at that time and place these women were all achievers, and that was the currency, that was the game that could be played, and by golly they went out and played that too."

The SETI enterprise extends well beyond the SETI Institute. Across the bay from the institute, at the University of California at Berkeley, astronomer Dan Werthimer and a group of colleagues run Project SERENDIP—the Search for Extraterrestrial Emission from Nearby Developed Intelligent Populations.[5] Begun by Jill Tarter and Stuart Bowyer in the late 1970s, Project SERENDIP was designed to get around the major impediment to SETI—the shortage of telescope time—by piggybacking multichannel receivers onto telescopes that were carrying out other studies. Thus, instead of a methodical, targeted search, Tarter and Bowyer simply got periodic deliveries of tapes containing narrowband data from whatever part of the sky the conventional radio astronomers had been observing. They began the program at Hat Creek Observatory in California.

Currently, however, Werthimer runs the SERENDIP observations at the much more powerful Arecibo telescope. This instrument offers unique advantages for the "parasitic" approach. Unlike most radio telescopes, the giant dish remains stationary, and targeting is done by moving the receiver. The "carriage house" that performs this movement supports an extra receiver system that does not compete with the primary receiver for focus. In fact, the secondary receiver can be moved independently, allowing some degree of independent targeting for SETI operations (Since our interview with Jill Tarter, she too has moved part of her operations to the Arecibo telescope).

The SERENDIP search at Arecibo is an all-sky survey: the sky is scanned in a raster pattern as the Earth rotates and the seasons progress. Each small patch of sky is studied only briefly, but repeatedly. The receiver analyzes 168 million frequency channels simultaneously, each 0.6 hertz wide. While much of the survey has been done near the 21-centimeter hydrogen line, Werthimer's group has also conducted a lengthy search at 70-centimeters, nearer to UHF television and radar frequencies. When we ask Werthimer, in a telephone interview, whether any candidate signals have been detected during the 6 years they have been at Arecibo, he answers with a cautious triple negative: "So far we've never had a signal that's not consistent with noise."

SERENDIP is supported by the SETI Institute, the Planetary Society (a society of astronomy buffs founded by Carl Sagan), and the Friends of SERENDIP, a fund-raising group run by Arthur C. Clarke. (Serendipitously, "Serendip" is an old name for the island of Sri Lanka, where Clarke lives.). Werthimer tells us that he is also planning to involve the public through "SETI@home," a program by which several thousand SETI aficionados will donate CPU time on their home PCS for the further analysis of SERENDIP data—looking for higher-order regularities, for example, than those routinely detected by the Berkeley signal-processing software.

Other long-running programs have been going on at Harvard University and Ohio State University. The Harvard project, run by physicist Paul Horowitz, uses a dedicated but less-than-ideal telescope in the town of Harvard, west of Boston.[6] Horowitz has always been a leader in the use of narrowband, multichannel receivers: his current device scans the "waterhole" with a 240-million-channel receiver, each channel being 0.5 hertz wide. The Ohio State program, an all-sky survey, was begun by John Kraus and Robert Dixon in 1973.[7] They used an unusual radio telescope, designed by Kraus, that featured a flat but tiltable primary reflector and a fixed, parabolic secondary reflector. The highlight of the Ohio State program was the famous "Wow" signal, detected in August 1977, which had every hallmark of an extraterrestrial signal but which never recurred after its one brief appearance. The program ran continuously until 1998, when the telescope was torn down to make way for a golf course.

There have also been significant SETI efforts outside the United States. France, Argentina, Italy, and Australia all host SETI searches of one kind or another. The Argentine program is an offshoot of Horowitz's program, while the Italian and Australian[8] programs derive from Project SERENDIP. The Australian and Argentine programs, of course, observe the part of the sky that cannot be seen from the United States.

Russia has little activity today, but it was a hotbed of SETI research during the Soviet era. Two astronomers at Moscow University, Iosif S. Shklovskii and Nicolai Kardashev, became interested in the existence of extraterrestrial civilizations during the early 1960s. Shklovskii wrote a book that was translated and expanded by Carl Sagan: it became the bible of SETI enthusiasts worldwide.[9] Among other provocative ideas, Shklovskii suggested that the two moons of Mars, named Phobos and Deimos, might be artificial space stations. Kardashev set his sights even higher. He proposed a three-level hierarchy of civilizations.[10] Type 1 civilizations, like our own, make use of the radiation of a star that impinges on their own planet; Type 2 civilizations harness the entirety of their star's radiation; and Type 3 civilizations harness the energy of their entire galaxy.

Kardashev suggested that these highly evolved civilizations, because of their colossal power resources and their need to transfer vast amounts of information, would transmit time-varying signals, with a very wide band of frequencies, centered on the "water hole." He named a couple of radio sources that might reward study; and in April 1965, an astronomer colleague at Moscow University announced that one of these sources, "CTA 102," did indeed have spectral and time-varying characteristics consistent with those predicted by Kardashev. CTA 102, the colleague indicated, might be a signal emitted by one of Kardashev's supercivilizations. The announcement attracted worldwide attention. Unbeknownst to the Russians, however, CTA 102 had just been identified by US astronomers as an extremely distant quasar.

The pro-SETI atmosphere in Soviet-era Russia may have been linked to that country's materialistic, antireligious state philosophy, in the same way as Lucretius's interest in other worlds went hand in hand with his disdain for religion. Even Jill Tarter expresses a hint of the same attitude, when we ask her about Ellie Arroway's conflicts with organized religion in Sagan's novel. "I had the same kinds of conflicts," she says. "You went in with questions, and logical answers were not being served up. You were just supposed to believe, and I was eventually asked not to come back."

Besides searching for extraterrestrial radio signals, SETI proponents have also sought ways to send signals in the reverse direction. In 1974, to mark the rededication of the Arecibo telescope after an upgrade, Frank Drake sent a 169-second message toward a cluster of stars known as M13, or the "Great Cluster of Hercules." The message, reconstructed as a raster, displayed some basic information about the solar system and a primitive sketch of a human and the DNA double helix.

The Nobel Prize winner and British astronomer Martin Ryle protested Drake's action, fearing some hostile reaction from the recipients of the message. Ryle's protest may have been overdone: it will take 25,000 years for the message to reach M13, and another 25,000 years for the M13 battle stations (unless they have warp drives) to reach Earth. There is even some question as to whether Drake correctly aimed the beam, given the expected proper (across-the-sky) motion of the M13 cluster during the next 25,000 years. But still, Ryle's objection raised serious issues about how and when humans should attempt to contact extraterrestrial civilizations. The current consensus is that, if a message from extraterrestrials is received, no reply should be sent without careful scrutiny and broad international agreement.[11] Nothing, of course, can be done about the television and other signals that have been unintentionally pumped into space for more than half a century.

Another symbolic gesture in which Drake and Carl Sagan were involved was the message plaques and phonograph records that were attached to the Pioneer 10 and 11 and Voyager 1 and 2 spacecraft. These spacecraft, launched in the 1970s, are now headed away from the solar system into interstellar space. The chances that any of these craft will come near a star in, say, the next million years are negligible. But the plaques served the useful function of stirring up public debate about how the human race should represent itself to aliens. Some critics claimed that, even though Drake and Sagan had done everything they could think of to help aliens understand what was on the plaques and the phonograph records, there remained insuperable problems of interpretation. Others didn't like the idea of full-frontal nudity being depicted on government property.

Within the growing community of researchers who call themselves astrobiologists, the SETI enterprise has fairly broad support. At a 1998 workshop at NASA Ames Research Center, whose purpose was to lay out a roadmap for NASA's astrobiology programs over the next couple of decades,[12] SETI was explicitly listed as an endeavor that NASA should support, even though it will take a change of heart on the part of Congress to make that a reality. But there are some scientists who consider SETI pretty much a waste of time. We meet with one of these "pessimists," astronomer Ben Zuckerman of UCLA, to hear the other side of the story.

"Completely crazy—impossible—literally nuts!" is how Zuckerman assesses Drake's estimate that there are ten thousand technological civilizations in the Milky Way galaxy. "There is no Drake Equation, that's the whole thing! The so-called Drake Equation is useful for describing the various things that would enter into what N might turn out to be, but the idea that you can calculate N with the Drake Equation is not where it's at—it's a sheer guess."

In his anti-SETI stance, Zuckerman displays something of the fervor of the recovered alcoholic preaching against the evils of drink. For he himself was once a SETI practitioner. In the early 1970s, when he was at the University of Maryland, Zuckerman and Pat Palmer of the University of Chicago searched about six hundred stars for alien signals, using telescopes at the Green Bank Observatory. They used a 380-channel receiver, tuned to frequencies near the 21-centimeter hydrogen line. They did pick up some "signals," but Zuckerman believes they were caused by interference.

Zuckerman now downplays the significance of his early engagement in the SETI enterprise. "I was never a true believer in the sense of a Drake or Sagan. I was sort of an agnostic. I was never fully convinced by their basically Copernican arguments—that we shouldn't be unusual or special. We could not be sitting having these discussions if everything hadn't worked out just right for us to be here, so I don't think our existence proves anything."

What most drove Zuckerman into the "pessimist" camp was the work of Michael Hart, described in Chapter 3. Hart calculated that the habitable zone around a star was vanishingly narrow. If the Earth were moved just a few percentage points closer to or away from the Sun, he argued, life could not have emerged and persisted. Although more recent work, such as that by Jim Kasting, seemed to expand the habitable zone considerably, Zuckerman isn't persuaded. He argues that Kasting misunderstands the relationship between atmospheric carbon dioxide concentration and planetary temperature. In the long term, he says, it's planetary temperature that sets the carbon dioxide concentration, rather than the reverse. So runaway greenhouse or runaway refrigerator effects are likely to be the rule, and climates compatible with liquid water are likely to be the very rare exception.

In fact, according to Zuckerman, there are other considerations that make habitable planets less probable than even Hart's analysis suggested. For one thing, according to some theorists, the development of life on Earth has been made possible by the presence of Jupiter, whose massive gravity has diverted large numbers of comets from the region of the Earth's orbit.[13] Without Jupiter, the argument goes, life would have been extinguished by giant impacts each time that it tried to gain a foothold. And given the findings on extrasolar planets that have been discovered so far, the arrangement in our own solar system may be highly unusual. "It may be that ninety-nine times of a hundred when you form a Jupiter it migrates inward and messes everything up in the inner solar system, and it's only very rarely that it's left out in the right place so it can protect the Earth from comets."

Another argument has to do with the Moon. Without a large, close moon, according to this argument, the orientation of a planet's spin axis would be unstable, and this would permit wild variations in mean temperature that would inevitably end up with runaway heating or cooling. How common are large, close moons around rocky planets? Zuckerman says that we should discount the example of Earth's moon, for "anti-Copernican" reasons. That leaves zero large moons around three rocky planets in our own solar system. "For all we know, it might be a one-in-a-million shot that we have a moon able to stabilize obliquity sufficiently to enable the climate to be stable." Thus, the term n_e in the Drake Equation—the number of habitable planets per star—may be so low as to drive down Drake's N by many orders of magnitude.

These arguments about the prevalence of habitable planets, involving as they do so many unknowns, seem more designed to inspire caution than to knock SETI off its foundations. But Zuckerman has another, much more interesting argument, which is specifically focused on the question of intelligent life in the universe. This argument has to do with the famous "Fermi Paradox."

The topic of extraterrestrial life came up over lunch at Los Alamos, in the summer of 1950. The nuclear physicist Enrico Fermi asked a group of colleagues, including Edward Teller, "If there are extraterrestrials, where are they?"[14] Although posed in the most casual of circumstances, this apparently simple question has reverberated through the decades and has at times threatened to destroy the credibility of the SETI enterprise.

One possible answer to Fermi's question is that extraterrestrials have in fact often visited the Earth, and continue to do so. This is the answer of the UFO community, of course (see Chapter 8). But few scientists, even those engaged in SETI, take the UFO claims seriously. "You won't find anyone around here who believes in UFOs," Drake tells us. Jill Tarter is only a trifle more open-minded on the issue: "I do believe that there are unexplained phenomena," she says. "I also believe that they have nothing whatsoever to do with little green men in flying saucers, because there's no tangible evidence to that end; and if you show me the evidence, I'll be happy to change my belief, because that's the difference between science and religion or fact and fiction—the difference is the evidence."

If one discounts the UFO claims, yet still believes that there are many technological civilizations in the galaxy, why have they not visited us? Drake's answer is straightforward: "High speed interstellar travel is so demanding of resources and so hazardous that intelligent civilizations don't attempt it." And why should they attempt it, when radio communication can supply all the information they might want?

At first glance, Drake's argument seems very persuasive. The distances between stars are truly immense. To get from Earth to the nearest star and back, traveling at 99 percent of the speed of light, would take 8 years. And Barney Oliver showed that, to accelerate a spacecraft to such a speed, to bring it to a stop, and to repeat the process in the reverse direction, would take almost unimaginable amounts of energy.

"Barney got it wrong in two ways," says Zuckerman. "One is that, of course, we're not going to travel at 99 percent of the speed of light; and second, once you get there, you don't come back. It's like when the Polynesians arrived in Hawaii: they didn't turn round and go back to wherever they came from, they stayed there. So his calculations are off by umpteen orders of magnitude."

Zuckerman's argument stands SETI on its head, because it has traditionally been the SETI folk who have been enthusiastic about interstellar travel. It was Drake, after all, who affixed the message plaques to the Pioneer and Voyager spacecraft that are already winging their way toward the stars, less than a lifetime from the beginning of the space age. And it is Drake who claims that technological civilizations can last a billion years. That's plenty of time, it would seem, to build larger spacecraft that could house migrants for however long it takes to find a hospitable planet—for hundreds of generations, if necessary.[15] Such migration would become a necessity for some very long-lived civilizations, Zuckerman has pointed out, because all stars eventually burn out.[16]

Even if the members of a distant civilization did not wish to travel in person, surely they would send robotic vehicles to explore the galaxy. This idea was put forward by Ronald Bracewell, the Stanford University astronomer who also first proposed applying optical interferometry to planet detection.[17] These "Bracewell probes" should be exploring and looking for signs of life in planetary systems, including our own. And physicist Frank Tipler took the argument even further.[18] The probes, he said, would be "cellular automata," machines capable of replicating themselves in the fashion originally laid out by mathematician John von Neumann. Even a few such self-replicating probes would eventually populate the entire galaxy, perhaps long after their inventors had become extinct.

Zuckerman pooh-poohs Drake's notion that technological beings would be satisfied with radio communication. "Drake's implicit assumption is that the only thing we're going to care about is intelligent life. But what if we have an interest in simpler life-forms? If you turn the picture around and you have some advanced extraterrestrials looking at the Earth, until the last hundred years there was no evidence of intelligent life, but for billions of years before that they could have deduced that this was a very unusual world and that there were probably living creatures on it. They would have had billions of years to come investigate." In short, the reason extraterrestrials haven't visited us is that they don't exist.

To Zuckerman, the words "optimist" and "pessimist," commonly used to describe the pro-SETI and anti-SETI camps, are misnomers. "Their view of what goes on in the universe is very unthrilling, in my opinion—that civilizations send radio signals back and forth, and that's the future of mankind. I think that the idea that there will be space travel by any civilization able to build rocket ships is a much more exciting picture of the future of mankind and the future of the universe. I really think that we're much more optimistic than those stay-at-home SETI types."

Because he believes that life—whether intelligent or not—is a rarity in the universe, Zuckerman also believes that we have a special responsibility to protect it. "It's just obscene the way human beings are wiping out half or more of the species on Earth in the course of the twentieth and twenty-first century. I don't think we're going to wipe ourselves out, but I do think that we're going to wipe out the majority of other species, and it's such a crime, given how rare I think life is in the universe."

Although Zuckerman is a "pessimist," he seems to have inched back toward a more open-minded position over the last few years. Whereas Michael Hart has argued that we are alone in the observable universe, Zuckerman is willing to believe that there may be one or two technological civilizations per galaxy. Spending money on SETI, he says, is worse than spending it on saving the rain forest, but better than spending it on nuclear bombs. And if the SETI search is privately funded, as is currently the case in the United States, he has no objections at all. "If rich people want to spend their money on this, I'm 100 percent in favor of it. What I do find remarkable though is that smart people are willing to devote a good part of their careers to searching for these signals. I can't imagine spending time on it anymore, given the views I've expressed to you."

Because Zuckerman's arguments depend so crucially on the feasibility of interstellar travel, we attended a NASA workshop at Caltech that was devoted to that very topic.[19] At the meeting, astronomers, engineers, and futurists discussed what might seem like an extremely modest goal: sending a small unmanned probe to the nearest star system, Alpha Centauri, at one-tenth the speed of light (0.1c). Such a trip would take 40 years (not including acceleration time). The probe would not return to Earth nor even decelerate when it reached its goal.

Although 0.1c is much slower and more economical than the 0.99c envisaged by Oliver, it is still ten thousand times faster than the speed at which the Voyager spacecraft are now heading away from the solar system. What kind of propulsion system could accelerate a spacecraft to such an extraordinary velocity? The meeting participants agreed that chemical rockets and even nuclear fission rockets could not do the job under any circumstances: they simply could not supply enough power for a given total mass of rocket, fuel, and spacecraft.

Instead, the participants considered a number of more exotic propulsion systems.[20] Steve Howe of Los Alamos discussed antimatter propulsion.[21] Antimatter hydrogen (an antiproton plus a positron) when "burned" with regular matter is the most energy-rich fuel known: it supplies about 10^{10} times the energy of the same weight of gasoline. But the world production of antimatter is currently less than 1 nanogram (a billionth of a gram) annually, whereas kilograms of the stuff would be required for an interstellar mission. There would also be extraordinary problems with storage of the fuel in a world where all containers must be made of regular matter.

Several participants discussed propulsion by nuclear fusion. Because fusion fuel, such as deuterium, is far less energy rich than antimatter, this kind of propulsion is extremely problematic. Fred Winterberg of the University of Nevada suggested that the spacecraft could save on fuel weight by stopping at comets in the Oort cloud (which may extend as much as halfway to the nearest star) and mining them for deuterium. Again, this idea seems speculative in the extreme, given how little we know about the Oort cloud and the problems of miniaturizing the mining and refinement equipment.

Because the mass of the fuel places such a burden on interstellar spacecraft, many people have considered propulsion systems in which the power is supplied by some kind of beam generated on or near Earth. It has been suggested, for example, that the spacecraft could ride a stream of pellets fired from a mass driver.[22] To accelerate the pellets to the required velocity, however, the mass driver would have to be 100,000 kilometers long!

At the Caltech meeting, the participants discussed other kinds of beams, such as beams of laser light. As originally formulated by futurist Bob Forward, the beam would strike a "light sail," driving it forward by the light's momentum in the same way as "solar sails," driven by sunlight, have been proposed for travel within the solar system.[23] Because the light from a laser will inevitably diverge with distance, however, at least one lens would have to be placed somewhere in space in order to refocus the beam. The lens would have to be at least a thousand kilometers in diameter—one-third the diameter of the Moon—and would have a mass of half a million tons, even if made of ultrathin plastic film. And powering the laser would require an estimated supply of 65,000 gigawatts of electricity for 900 hours: at current US rates, this amount of electricity would cost about $2 trillion, plus as much as $130 trillion to construct the electricity generation facility![24]

One participant at the meeting, Leik Myrabo of Rensselaer Polytechnic Institute, is way ahead of his rivals: he has actually constructed a prototype of his spacecraft and launched it from Earth.[25] A grapefruit-sized disk, it was powered by a beam generated by an ultrapowerful laser. The laser energy heated the air under the disk, propelling it upward to a height of 90 feet—fully one-quadrillionth of the distance to Alpha Centauri. Then it crashed in a heap of scorched titanium.

Not quite so far along is Bernie Haisch, of Lockheed-Martin.[26] He is investigating the idea of accelerating spacecraft by tapping into the quantum vacuum—the space from which, according to theory, pairs of virtual particles are constantly appearing and disappearing. As a source of motive power, however, the quantum vacuum is presently about on a par with the "force field" beloved of science fiction writers.

The participants in the workshop agreed on only one thing: that it would take "miracles" for any of their schemes to work. Of course, as Zuckerman hastens to points out, a jetliner would be a miracle to Christopher Columbus. It may be that everything will fall into place and the mission to Alpha Centauri will begin on its projected launch date (the year 2040).

But Zuckerman's case against SETI requires, not that interstellar travel *may* be possible, but that it *is* possible—and not merely possible but actually preferable to other forms of communication that seem a whole lot easier and cheaper. That, surely, has not yet been demonstrated. To believe strongly in Zuckerman's case seems to require believing in the indefinite continuation of exponential processes such as the drop in energy costs, the increase in scientific discoveries, the replication of von Neumann machines, and so on. In the real world, however, something eventually slows exponential processes down.

Jill Tarter is unfazed by any of the arguments that the "pessimists" have raised. The Fermi Paradox she rejects for an even more basic reason than the one offered by Drake. "I object to the whole logical construct," she tells us, "because in order for it to make any sense, you have to take as fact that extraterrestrials are not here. Now I don't think that they're picking up Aunt Harriet and Uncle Charlie off the streets and taking them up to their spacecraft for nefarious purposes. But we can't say for sure that there isn't some long, slow spacecraft orbiting the asteroids and chopping up a bunch of raw materials, which some people suggest that an advanced civilization might do.[27] We don't have a General Accounting Office that counts asteroids and says, 'Oh, we lost three this week, there must be someone out there.' So I don't even agree that they're not here."

Tarter's statement surprises us. It seems to come perilously close to a belief in UFOS, even though she denies such a belief. After all, if aliens are indeed in our solar system, they surely would be tempted to visit Earth and abduct a few of us, whether for scientific purposes or merely to restock their freezers. Because we are extremely skeptical that aliens have ever landed on our planet, we are also extremely skeptical that aliens lurk anywhere in our solar system.

After listening to all the arguments of the "optimists" and the "pessimists," we can't help but focus on the magnitude of our ignorance. Trying to derive real numbers from the Drake Equation may well be a sterile exercise, but the Fermi Paradox and the other arguments against SETI also seem too contrived and indirect to make a decisive case. Biology is, after all, an empirical science in which theoretical predictions have contributed relatively little to the advancement of knowledge. When we suggest to Tarter that theory is not going to answer the question of extraterrestrial intelligence, her response is heartfelt. "No!" she almost shouts. "And that's why I say, let's try an experiment."

SUGGESTED READING

Drake, F., and D. Sobel. *Is Anyone Out There? The Scientific Search for Extraterrestrial Intelligence.* New York: Delacorte Press, 1992.

Shostak, S. *Sharing the Universe: Perspectives on Extraterrestrial Life.* Berkeley: Berkeley Hills Books, 1998.

Zuckerman, B., and M.H. Hart. *Extraterrestrials: Where Are They?* 2d ed. Cambridge: Cambridge University Press, 1995.

8

Dreamland

The Science and Religion of UFOs

After a night spent among flashing neon marquees and chattering slot machines, the trip north from Las Vegas seems like an exercise in sensory deprivation. There's little traffic on the freeway, even less on State Highway 93, and none at all after we turn onto Highway 375 toward Rachel, Nevada. Perhaps it's the official green sign that scares people away. Posted at the turnoff, under the last tree that we'll see for a hundred miles, it reads "Extraterrestrial Highway." And for those who don't understand English, there's a sketch of some angular, improbable flying machine—a Stealth version of a UFO, perhaps.

The road takes us through classic basin-and-range country. We're traveling north along one of the Earth's stretch marks, a gently sloping alluvial basin between two mountain ranges. The ranges are giant blocks of the Earth's crust that cracked and keeled over, 15 million years ago, when California and Utah started a tug-of-war for possession of Nevada. The geological drama continues today, but at too slow a pace to hold our attention. In fact, there's nothing to look at but sagebrush, the long ribbon of road, and the empty, silent sky. At least, among these limitless vistas, no alien could sneak up on us unawares.

After 20 miles, an object finally comes into view. It's a white mailbox, marking the start of a dirt road that heads over toward the hills in the west. We stop to inspect the queer little token of civilization. With rising excitement we realize that this mailbox, in spite of its color, is none other than the infamous "Black Mailbox"—quite possibly the locale of more UFO sightings than any other spot on Earth. The box's owner, rancher Steve Medlin, has evidently replaced the old mailbox with a larger, lockable box and has painted it white to throw off UFO aficionados. But they are not to be thrown off that easily. In fact, the white-over-black finish has been a boon to them, because it offers a fine medium for intaglio-style graffiti. Scratched into the surface are the names of visitors from the four corners of the Earth, if not beyond. And alongside the names are brief messages that tell of the motive for their pilgrimages, such as "Waiting for the Mothership" or, more simply and poignantly, "I Believe."

Standing by the mailbox, we scan the horizon—eagerly at first, then with growing impatience. We feel like Cary Grant, after he got off the Greyhound bus in *North by Northwest*. But here, not only are there no crops, there is no crop duster either. And there's no *Aurora*, the Mach-8 spy plane that's being tested, we're told, at "Area 51," the supersecret military base to the west of us. And most definitely no UFOs in the prohibited airspace above the base, an airspace that military pilots call "Dreamland." Why, indeed, should UFOs show up for the likes of us, a couple of skeptics who would likely dismiss them as optical illusions, even if they landed 20 feet from us and discharged a platoon of bug-eyed monsters?

Oppressed by the silence and emptiness of the landscape, we return to our car and continue northward. After 10 miles, we pass a dead cow. Another 10, and we reach the "town" of Rachel—a scattering of trailers that, had we been looking down to fiddle with the radio, we might well have missed. Yet a Michelin guide to central Nevada would probably give Rachel at least a single star. That's on account of two trailers, at the opposite ends of town, whose owners represent two warring factions within ufology.

The first trailer is painted a bilious yellow and boasts a large sign saying "Area 51 Research Center." The sign is a bit outdated. Glenn Campbell, the trailer's owner, has moved his research operations to his new home in Las Vegas. There he runs "Ufomind," a website where you can find out all that you ever wanted to know, and a whole lot more, about UFOs and the paranormal.[1] His trailer at Rachel now sells books and UFO collectibles, and we politely purchase an Area 51 coffee mug.

"When I first moved out to Rachel from Boston in '92," Campbell tells us, "it was because of the UFO claims—the specific claim that you could see flying saucers on demand. You just go to the black mailbox any Wednesday night. That was a glorious thing to check out, because it's either true or false.

"The first night, I didn't see anything. But the second night, I saw these incredible golden orbs, brighter than any star or planet, almost the brightness of the moon, down in the general direction that they're supposed to be, which is the direction of Papoose Lake [a dry lake bed located deep within the prohibited military area]. These things would simply appear above the horizon and hover for a while and slowly drift from one side to another and then vanish somehow. They matched some of the descriptions I'd read in the UFO magazines, so I thought, 'God, I've seen a UFO!' But it was kind of a disappointing UFO—it didn't make these stair-step motions or these incredible turn-on-a-dime things that they're supposed to.

"The next night, I went to a different location, near Rachel, and saw them again. With my binoculars I could see little strobe lights around them, as if the UFOs were accompanied by helicopters. So I did a triangulation: I took a compass reading of where they rose, and the third night, I went 40 miles down the road and took another reading and plotted the directions on a map. This told me that these 'flying saucers' were not at Papoose Lake, and not in Area 51 at all, but in the bombing ranges well to the south.

"That third night, they were so bright, like great glowing orbs, that they lit up the ground below through the cloud cover. And coming from each of them, I saw a plume of black smoke rising. So I had to recognize that these flying saucers were probably petrochemically powered. And that's when it occurred to me that they were probably flares, and this was borne out later. So in a week of investigation, I'd pretty much debunked them, although that wasn't my intention. I'd come to see whatever the truth might be."

Yet the Area 51 mystery wasn't solved as easily as that. Three years before Campbell arrived in Rachel, a colorful character by the name of Bob Lazar had caused a major stir within the UFO community. A self-described "physicist" who had previously been employed at the Los Alamos National Laboratory, Lazar was interviewed about his experiences at Area 51 on a Las Vegas television show. Earlier that year, he said, he had been hired to work on some "flying disks" that had been captured from extraterrestrials. They were being studied at Papoose Lake, a few miles to the south of the main Area 51 facility at Groom Lake.

Except for one brief glimpse of a "small, grey alien," Lazar didn't meet the extraterrestrials themselves. But what really got people excited was Lazar's description of the machines, especially their gravity drives. Unlike the familiar gravity-canceling drives that provide lift for many alien craft (such as the Martian vehicles in *Auf Zwei Planeten*), these drives actually amplified gravity. Centered on plates made of "Element 115" that emitted a stream of antimatter particles, they warped space-time in the direction ahead of the vehicle, thus permitting instantaneous travel across the galaxy. Military pilots, Lazar said, were already test-flying the disks in the Papoose Lake area.

It was Lazar's story that really put Area 51 on the UFO map. For a while he was the center of attention, as floods of UFO seekers traveled Highway 375 in search of the amazing craft. Often the pilgrims did see UFOs, or thought they did. But gradually, Lazar's credibility began to wane. It probably didn't help that, a few months after his television interview, he was arrested and convicted on a pandering charge. Or that several attempts to verify his claimed master's degrees (from MIT and Caltech) were unsuccessful. Or that his descriptions of facilities at Area 51 conflicted wildly with the recollections of people who had worked there.

"There's a certain event cycle in the UFO field," says Campbell. "A claim will be made, there's a big hysteria about it, and then the skeptics come in and find flaws with the claim, and it expires. The cycle had pretty much completed for Bob Lazar by the time I arrived on the scene, but I still had an open mind about it, and plenty of other serious people still believed him." So Campbell began a 2-year-long study of Area 51, involving frequent observation of the Groom Lake base from nearby mountaintops (partly because of Campbell's activities, these viewpoints have since been included within the prohibited area). The base and its environs have also been well documented in Russian spy-satellite photos, which are now widely available. Some intrepid souls have physically penetrated into the forbidden zone: in 1998, for example, an archaeologist by the name of Jerry Freeman actually reached Papoose Lake, but saw nothing unusual.

"At the beginning, it was a big terra incognita," says Campbell. "But in the intervening time we've found out what most of those hangars are there for. I can't say there's no aliens out there, but the box in which they might be housed is getting smaller and smaller."

What really ruined the mystique of Area 51's UFOs was *Independence Day*, which was partly located (but not shot) at the secret base. The opening of the film, in April 1996, was marked by a gala Hollywood party, held at Rachel. A plaque was unveiled, and Nevada's governor dedicated the Extraterrestrial Highway. From then on commercialism took over. At stores along Highway 93, for example, one can buy ready-to-autopsy plastic aliens, their stomachs already partly slit open to reveal some nasty looking bowel contents.

At Rachel, however, only two establishments are in a position to profit from the UFO seekers. One is Campbell's bookstore, and the other is Campbell's rival and nemesis, the "Little A'Le'Inn" (pronounced "Little Alien"). This trailer is a bar, eatery, and bare-bones motel, run by a couple named Pat and Joe Travis. Whereas Campbell is fortyish, balding, trim, and an obviously well-educated urbanite, the Travises are considerably older and look and act like rural survivalists. Joe, who runs the bar, sports a Ted Kaczynski beard and hairstyle. He is never without a beer, a gun, and a bad word for the federal government.

The Travises are absolutely and unreservedly proalien. The walls are covered with photos of UFOs, including the glowing orbs as seen from the Black Mailbox. There are also satellite images and maps of Area 51. UFOs are sighted "all the time," they say. The *Independence Day* plaque is cemented into a boulder immediately outside the door of the bar—a location that clearly maddens Campbell, who was thrown off the Travis's property in 1993 and has never been back. During the unveiling, Campbell actually staged a counterdemonstration accusing the film's producers, the Governor of Nevada, and the Travises of assorted acts of bad faith.

To people like Joe Travis, Campbell is nothing but a carping critic of ufology, or possibly a US and/or Russian agent planted to sow disinformation about extraterrestrials and Area 51. The two men, originally good friends, now badmouth each other in their separate fiefdoms: Travis at his bar, Campbell in cyberspace.

Yet Campbell is not an unbeliever. "I'm impressed by some of the early accounts, in the 1950s and 1960s," he says. "It was just sightings, it wasn't mucked up with the abduction stuff. A typical sighting was: Someone's driving down the road, and suddenly there's a light in the sky, and the car conks out, and the radio goes 'woo-woo-woo.' And people with no real contact with each other would report the same physical effects. There's a reason that the radio would go 'woo-woo-woo' because that's what a radio does in the presence of a magnetic field."

The modern UFO era actually began in 1947, when a civilian pilot reported seeing nine disklike flying objects near Mt. Rainier in Washington State. Within months, hundreds of sightings had been reported from around the country. On January 7 of the following year, a UFO (probably a balloon) was seen by a large number of people in Kentucky, including military personnel at Godman Air Force Base. A plane dispatched to investigate the UFO crashed, killing the pilot. This sparked the first of a series of investigations of the UFO phenomenon by the US Air Force: Project Sign (1947), Project Grudge (1947–49), and Project Blue Book (1952–69).[2] The report of Project Sign apparently concluded that the UFOs were alien spacecraft, but the report was rejected and destroyed by the Air Force itself. This, of course, was a major boost to the notion—now an article of faith in the UFO community—that the US Government knows a lot more about UFOs than it is willing to let on.

The Air Force's most notable contribution to the study of UFOs, however, was its commission of the Condon Report, released in 1969.[3] Edward Condon, a physicist at the University of Colorado who had earlier been Director of the National Bureau of Standards, led a 2-year-long, half-million-dollar investigation of the UFO phenomenon, building on the long-running Project Sign study. The investigation included not merely interviews with witnesses and examination of photographs but also laboratory tests, such as tests for remanent (residual) magnetism in vehicles that had been approached by UFOs. Condon summarized the nearly thousand pages of the report by writing: "Our general conclusion is that nothing has come from the study of UFOs in the past 21 years that has added to scientific knowledge." And he added: "Further extensive study of UFOs probably cannot be justified in the expectation that science will be advanced thereby."

Although responses to the Condon Report were predictably diverse, its general effect among scientists was to dampen interest in the topic of UFOs, especially with regard to the hypothesis that UFOs were alien spacecraft. In the wider society, however, interest in UFOs continued unabated, and the UFO movement became progressively zanier. Abductions became commonplace. Aliens brought new healing powers, but the medical profession was suppressing information about them. Evidence of a giant government cover-up, reaching back to the notorious UFO crash at Roswell, New Mexico, in 1947, was set forth in numerous books. And the German writer Erich von Däniken began a long-running series of popular books devoted to what one might call "paleo-ufology": the notion that alien technology was used to build the pyramids of Egypt and other ancient landmarks. His first book, *Chariots of the Gods?* had one redeeming feature—the question mark at the end of the title—but the following volumes, such as *Miracles of the Gods* and *Gold of the Gods,* dispensed even with that.

Over the last 30 years, a new breed of professional ufologists has appeared, some of whom have respectable scientific credentials. One of these is the French-born Jacques Vallee, who has a PhD in computer science from Northwestern University. Like Jill Tarter, Vallee has been immortalized by Hollywood: he is the model for the French ufologist, played by François Truffaut, in the film *Close Encounters of the Third Kind.* Vallee, who now lives in San Francisco, has led ufology away from space-faring aliens and gravity drives and toward more metaphysical concepts. UFOS, he suggests, could be the multidimensional "fractal" emanations of terrestrial but nonhuman consciousness—whatever that might mean.

Another person with academic credentials who has recently made waves in the UFO world is the Harvard Medical School psychiatrist and hypnotist John Mack.[4] A number of Mack's subjects, while under hypnosis, described having been abducted by aliens. In a 1995 book, Mack declared that he believed his subjects. This, in spite of the fact that most of the subjects displayed classic signs of fantasy proneness, such as having had imaginary playmates during childhood.

In 1998, a new UFO study was published.[5] This one was carried out under the auspices of the Society for Scientific Exploration, a group interested in all kinds of "anomalous" phenomena such as dowsing (water divination), cryptozoology (the Loch Ness monster and her ilk), and telepathy. The society's founder, Stanford University physicist Peter Sturrock, directed the study. In it, a panel of nine physical scientists listened to presentations by eight ufologists over a long weekend. The latter included Jacques Vallee, who publishes some of his research in the society's journal.

Like the Condon Report, the report of Sturrock's panel concluded that "there was no convincing evidence pointing to unknown physical processes or to the involvement of extraterrestrial intelligence."

Nevertheless, the report repeatedly used terms like "intriguing" to describe the ufologists' presentations and, in a sharp departure from Condon's conclusions, recommended that UFO studies continue and that institutional support be provided for them.

We talk by phone with Von Eshleman, a retired Stanford physicist who was one of the panel's cochairs. Eshleman tells us that he previously had no special interest in UFOs. "I got a cold call from someone I knew, and it sounded like an interesting subject, and I'm emeritus, so I don't have to worry about my colleagues saying I shouldn't be doing these things. The concept of not even listening to the reports I think is a mistake, and there are examples in the scientific literature where scientists have been wrong for very long periods of time. An example is continental drift—that was ridiculed by the experts for decades after it was first proposed [by German meteorologist Alfred Wegener].

"There was an example of the same thing at our meeting, concerning lights observed in Norway. When I heard the presentation by the UFO investigators, I realized that they were due to electrical activity above thunderstorms. This is something that has just recently been accepted by experts as occurring—previously, the experts had said that all of the electrical activity of thunderstorms takes place in or below the clouds. Even the UFO investigators had not heard of this new turn in the scientific literature."

The idea that following up reports of UFO sightings might cast light on previously unknown natural phenomena seems reasonable enough. But, of course, the public debate is about much more than that. "UFOs remain unidentified because there isn't enough evidence to go beyond the unidentified category," Eshleman says. "Unfortunately many people, when you say UFO, think that that means a visitation of aliens or a government cover-up or something like that."

Certainly, the ufologists who presented at the meeting were making a pitch for more than natural phenomena. Richard Haines of Los Altos, California, presented a photograph taken by a woman in British Columbia, which included a definitely artificial-looking object shaped like a flying saucer in the field of view. (The woman said that she had not even seen the object when she took the photograph.) Michael Swords of Western Michigan University recounted a well-known UFO sighting from 1973 in which the crew of a US Army Reserve helicopter flying near Mansfield, Ohio, saw a grey, metallic, cigar-shaped craft that hovered above them and bathed them in green light. The UFO, Swords said, was seen independently by a family in a car on the ground. And Jacques Vallee recounted an event that took place in Council Bluffs, Iowa, in 1977, when a mass of molten metal appeared mysteriously in a city park.

"That kind of situation is asking for hoaxes," says Eshleman of the Iowa case. "There have been examples where it was a similar mixture as [that made by] a foundry in the same town, and certainly not compounds or elements or states of matter that are unworldly."

"I thought we wrote a pretty mild report," he goes on. "We only changed the Condon Report in a minor way. We agree that nothing so far has really overturned science, but let's keep an open mind. I was surprised the media picked it up the way it did, but for a scientific group to say anything that doesn't completely debunk the UFOs, I suppose, was a surprise."

For a more critical voice, we turn to Philip Klass, a retired senior editor of *Aviation Week and Space Technology* and a long-time debunker of UFO sightings. "There's nothing that I would like better than to be able to write a story saying that at last I've investigated a UFO incident that has no prosaic explanation; but in 32 years, I have yet to find one. Sure, Eshleman says we should keep an open mind, but about what? Should we keep an open mind that maybe the Earth is flat and that Jimmy Hoffa fell off it? That the Bible is literally true and the dinosaurs were alive just a thousand years ago? Where's the evidence? All it would take would be one authentic photo. That photograph of the flying saucer in British Columbia that Dr. Haines showed—is it conceivable that a UFO should so closely resemble a Frisbee, and is it a coincidence that her husband enjoyed Frisbee throwing?"

Klass is even more critical of the Ohio report, in which the helicopter crew spotted an unidentified cigar-shaped craft. Klass spent many weeks investigating that incident and interviewed the pilot several times. "I'm satisfied that it was a meteor fall," Klass says. Although Swords, the ufologist, told the panel that there were unusual physical effects during the sighting, such as the failure of the helicopter's radio and compass, Klass isn't impressed. During the interviews he had with the pilot shortly after the incident, Klass says, the pilot never mentioned the compass, and he agreed that his difficulty in making radio contact with his home base was probably due to his low altitude at the time. In fact, at Klass's suggestion, the pilot returned to the same spot in another helicopter and again failed to make radio contact. Klass says that the pilot embroidered his story months after the incident, when he became a national UFO celebrity. As for the witnesses on the ground, Klass points out that they did not come forward until 3 years after the incident, when all the details of the sighting had been widely publicized. Yet none of these facts, he says, were made available to the panelists.

Klass criticizes the study's director, Peter Sturrock, for failing to include any skeptical voices among the UFO experts who presented material to the panel. And Klass even offers a theory for why intelligent, well-educated people like Sturrock end up promoting aliens. "I have a theory, based on a limited number of examples, that some scientists, when they get along in years and do not achieve great things, unconsciously reach out for fame by adopting far-out ideas. And I include John Mack in that. He wrote a Pulitzer Prize–winning book a long time ago; but now he's getting on in years, and he hasn't achieved fame except through his alien-abduction hypothesis." Klass ridicules Mack's notion of the aliens as consciousness-raising creatures. "His extraterrestrials tell their abductees, 'Don't engage in nuclear warfare, and be concerned about the ozone layer.' As if you might think that nuclear warfare would be good for our planet, if you hadn't been abducted!"

For yet another viewpoint, we speak with Paul Kurtz, a retired professor of philosophy at the State University of New York at Buffalo. Kurtz has been for many years one of the leading voices of secular humanism in the United States, and he is also chairman of the Committee for Scientific Investigation of Claims of the Paranormal (CSICOP), with which Klass is also affiliated.[6] "I consider ufology a new religion or quasireligion," Kurtz tells us. "It was spawned in the last 50 years, so we've been able to see it develop before our eyes. It's a psychological or sociological phenomenon, essentially—or neurological: it's in the eye of the beholder.

"The parallels between classical religions and the UFO mythology are very close. These creatures from outer space descend upon the Earth, they're more intelligent than us, and some of them are good and some are malevolent. They open up new dimensions of reality above and beyond our own. Usually you have an effusion of light; and the people that are beamed up or abducted, as in a mystical experience, are brought to another realm." (Another philosopher, Paul Davies, has also discussed the analogies between alien contact and religious experience.[7])

When we refer to CSICOP's role as "debunking," Kurtz objects. "We're not debunking. I don't like that description. Our duty is to keep an open mind and look at any responsible claims. But in the current cultural milieu, most of the reports are pro-UFO, and the large part of the culture accepts them; so when you find the truth about these claims, you have to make it known to the public. The most skeptical part of society is the scientists, because a lot of us have sifted through the reports and seen a psychological or sociological explanation. We think that it's possible that there's life in other parts of the universe, and we support SETI, but that doesn't translate into proof that we've had visitations."

If religion and UFOS offer something that people need, we ask Kurtz, what can he offer in their place? "The truth," he says. "The search for true knowledge is just as mysterious and intriguing. To probe the origins of our solar system and our own origins—that's very exciting to me."

One thing is clear, that opinions about UFOS, like beliefs about religion, are strongly held and not easy to shift. As Eshelman tells us, "Things are so polarized that it's very difficult to say anything sensible between the two extremes that will have any effect. I think the reports will continue, and the one group will continue to debunk, and the other group will continue to say there's something there and the government is covering it up—all that's going to continue. Maybe the aliens, when they finally arrive, will put a stop to it."

From within the UFO community, Glenn Campbell probably represents something closest to a conciliatory view. "In the UFO field, you've taken away all the rules. Anything is possible. And what you see in those circumstances is not what's out there but what's inside the mind of the observer. If you ask me, Do I believe we're being visited by aliens? I'd have to give my standard answer: I don't know. If there are aliens, that's certainly interesting to me, but I can keep going even if there turns out not to be aliens. I'm interested in why we're here, the meaning of life—those sorts of questions—and UFOS are a cutting-edge place to study them."

Leaving Rachel, we continue on in a vast loop across the sawtooth terrain of southwestern Nevada and back into the Owens Valley of California, heading home for Los Angeles. As we drive by the line of giant radio telescopes at Big Pine, where double-jointed aliens finally closed in on Charlie Sheen in *The Arrival*, we stop to stretch our legs. The immense bulk of the Sierra Nevada range is outlined against the last glimmer of dusk. Suddenly, to the south, a brilliant ball of light appears in the sky. Seemingly just a few miles ahead of us, it sinks calmly and silently to the horizon and disappears.

Our very own UFO? Or the product of brains besotted with talk of them? We drive home in a kind of chastened silence. But when we turn on the 11 PM news, we learn that Southern California has indeed received an extraterrestrial visitor: a meteor fall, somewhere to the west of Los Angeles.

SUGGESTED READING

Peebles, C. *Watch the Skies! A Chronicle of the Flying Saucer Myth.* Washington, DC: Smithsonian Institution Press, 1994.

9

Exotica

Life as We Don't Know It

The Field and Mobile Robotics Building on the campus of Pittsburgh's Carnegie-Mellon University is hard to find and even harder to get into. Built on steeply sloping ground, hemmed in by construction sites, and partially embedded within other buildings with confusingly similar names and functions, it offers such a challenge to our navigational skills that we end up, 10 minutes late, on the roof of an adjacent building from which we clamber by a kind of iron skywalk to the office of roboticist, futurologist, and general out-of-the-box thinker Hans Moravec.

The difficulty of getting into the building has a countervailing benefit: it makes it hard for the robots to get out. Most of them, in fact, sit morosely in a cavernous, hangarlike space, in various stages of development or decrepitude. Some may have recollections of the great outdoors, if their memory chips haven't been cannibalized for later models. "This one drove from Washington, DC to San Diego, 98.5 percent of the time automatically controlled," Moravec tells us, waving his hand toward what looks for all the world like a Plymouth minivan. "The earliest one was a big truck, crammed with mainframe computers and air-conditioning and power supplies. The drive train was converted to hydraulics so it could travel slowly enough. Then we went to computer workstations, and that's what this Humvee carried." He points to what must once have been a military ambulance: a red cross still graces its roof. "The latest vehicles are just regular cars. You clip a small camera on the rearview mirror, facing forward, and the whole thing is run by a lap-top sitting on the passenger seat." Buckled in, hopefully.

The road-navigating robots are a little too carlike to evoke any particular feelings of awe or empathy. But Moravec takes us into his own, smaller lab, which houses a family of waist-high machines resembling R2-D2, the lovable robot in George Lucas's 1977 movie *Star Wars*. "This one is Uranus, our most recent," he says proudly, introducing us to a creature with three eyes, a skirt of ultrasonic range finders, and wheels within wheels that allow it to move smoothly in any direction. "At the moment, its computer is only about 50 MIPS [million instructions per second]. We need about 1000 MIPS, but that's too large to fit on board. It may have to tow its brain behind on a trolley.... Over there is Neptune, the previous one, which we just brought back after many years of absence. Pluto is in a computer museum in Boston. It never really worked properly."

Pluto, Neptune, and Uranus are all predecessors to a multipurpose domestic robot, roughly reptilian in intelligence, that Moravec hopes to develop and commercialize within a few years. That, in turn, will be followed by a 100,000-MIPS, mouse-level machine, predicted to evolve by the year 2020. Ten years later, it will be superseded by a 3,000,000-MIPS, monkey-level machine, capable of learning by internal simulation of the world. By 2040, there will be 100,000,000-MIPS machines, equal in computing power to the human brain and capable of abstract reasoning. They will love humans, for a while at least, Moravec tells us reassuringly. They'll be our children, after all.

So are robots alive? Are they conscious? Or will they ever be? Moravec has strong views on these questions, which we'll get to eventually. But first, we need to bridge the gap between the drab Life we're familiar with and the colorful fringes of bio-fantasy, where Moravec is most at home.

In earlier chapters, we mentioned the possibility of various modest variations on terrestrial-style life. We discussed, for example, whether different versions of the genetic code might serve as well as the one in use on Earth. We tentatively identified reasons that life might prefer the code that is used on Earth over many alternative codes, but still, no one would expect to find the identical code operating on an alien planet, unless that Life and ours had a common ancestry.

We also described some of the evidence that our current DNA-RNA-protein world evolved from a chemically simpler world, based on RNA or RNA analogs (if we believe Leslie Orgel) or on peptides (if we believe Stuart Kauffman). These worlds, and their even simpler predecessors, could presumably be the basis for Lifes currently existing on planets elsewhere in the cosmos. After all, life on those planets might have got going more recently than it did here, or circumstances might have prevented evolution from an RNA or peptide world toward a DNA-RNA-protein world.

Another alternative biochemistry that we've already mentioned would be one based on a chirality (molecular handedness) opposite from that of our own Life—on right-handed rather than left-handed amino acids, for example. As we discussed, some evidence suggests that the relative abundance of left- and right-handed amino acids may be influenced by the circular polarization of radiation in the molecular clouds where stars are born. If so, star systems born in some regions of the galaxy might contain living systems based on left-handed amino acids, while those born in other regions might house organisms based on right-handed amino acids. Because stars, over their lifetimes, wander far from their birthplaces, examples of these two life-forms would eventually become intermixed in the galaxy, so that a Life based on right-handed amino acids might exist not too far from our own solar system.

Yet to anyone familiar with science fiction or futurology, all these variations on terrestrial Life will seem quite pedestrian. For humans have succeeded in envisaging Lifes that owe nothing to carbon-chemistry, nothing to chemistry of any kind, nothing to evolution, and nothing even to Nature. Some people even claim to have created fundamentally novel life forms in the laboratory. Exploring these exotic Lifes may sometimes strain our credulity, but doing so may help us think more clearly about what life really is.

If we want to abandon carbon, but to remain within the domain of chemistry, the challenge is to find other elements that emulate carbon's protean nature: its ability to form a limitless variety of active molecules, from simple starter materials like carbon dioxide and hydrogen cyanide to the mighty polymers that store energy, impose structure on protoplasm, catalyze complex reactions, and carry genetic information. Among the candidates to take carbon's role, one has been singled out more frequently than any other: silicon.

Silicon sits right below carbon in the column of the periodic table known as Group 4 (or Subgroup 14 by more recent convention). Below silicon come germanium, tin, and lead. What all these elements have in common is that their outer electron shells, which are capable of holding eight electrons, actually contain only four. Atoms of Group 4 elements tend to fill those four "empty slots" by sharing pairs of electrons with other atoms. In the compound methane (CH_4), for example, the carbon atom shares electron pairs with four hydrogen atoms: each shared electron pair belongs simultaneously to a hydrogen atom and to the outer shell of the carbon atom, which is thus filled. In carbon dioxide (CO_2), the carbon atom shares two electron pairs with each of two oxygen atoms, again filling its outer shell. In hydrogen cyanide (HCN), the carbon atom shares three electron pairs with the nitrogen atom, and one pair with the hydrogen atom, once again filling its outer shell. These electron-sharing arrangements are called covalent bonds (single, double, or triple), and they're the main reason that organic molecules and organic creatures don't fly apart.

Given the similarity between carbon and silicon, it's tempting to think that silicon could form the basis of an alternative biochemistry, every bit the equal of carbon chemistry. Silicon, after all, is reasonably abundant, especially in the crust of rocky planets like Earth. The other Group 4 elements are much less abundant, and for that reason alone would be poor ingredients for life.

To get an informed opinion on this matter, we speak with Claude Yoder, who is a professor of chemistry at Franklin and Marshall College in Lancaster, Pennsylvania. Yoder has spent years trying to make silicon behave like carbon, with surprisingly limited success. "In my opinion, the likelihood of a biochemistry based on silicon is very low," he tells us. "Carbon forms many double bonds and triple bonds to other elements, and these bonds play an important role in many biological molecules: amino acids have double bonds between carbon and oxygen, for example. But silicon is very reluctant to form double bonds. In fact, for many years it was thought that silicon did not form double bonds to other elements at all."

This difference in behavior has striking consequences, even at the level of very simple molecules. Both carbon and silicon form dioxides, for example; but carbon dioxide (CO_2) is a reactive gas, whereas silicon dioxide (SiO_2) forms an inert mineral, silica, that is a major constituent of terrestrial rocks. The reason for the difference, Yoder tells us, is that a carbon atom forms double bonds with two oxygen atoms, while a silicon atom prefers to form single bonds with four oxygen atoms. Each of those four oxygen atoms still has a vacant slot in its outer electron shell, so it readily bonds to another silicon atom, and that silicon atom bonds to other oxygen atoms, and so on. Thus, silicon dioxide ends up forming a tough, unreactive polymer whose formula is $(SiO_2)_n$.

How can two atoms in the same group of the periodic table behave so differently? The main reason is that the outer electron shell of the silicon atom is farther from the nucleus than is the case with carbon, so that the negative charge of the electrons in that shell is less effectively neutralized by the positive charge of the nuclear protons. Therefore, charge plays a greater role in the bonds formed by silicon than in those formed by carbon (in technical terms, the "percent ionic character" of the bonds is greater). This increases the strength of the single covalent bond. "It's not that a double bond between silicon and oxygen is inherently weak," Yoder says. "In fact, there's some evidence that it's about as strong as the double bond between carbon and oxygen. But the single bond is much stronger, so that's what you nearly always get."

Another problem with silicon is its reluctance to form long chains. "Carbon forms enormously long chains, in substances like Teflon," Yoder says, "in which one carbon is bonded to another in polymeric form. Silicon does this to a limited extent, but nowhere near the extent to which carbon does. This, plus its reluctance to form double bonds, almost precludes a biochemistry based on silicon."

Not all organic molecules are built of continuous carbon chains, however. In proteins, for example, the molecular backbone is made of carbon atoms interspersed with nitrogen (... CCNCCN ...), while in nucleic acids the carbon atoms are interspersed with oxygen and phosphorus atoms (... OPOCCCOPOCCC ...). Could silicon form macromolecules of this type? "Yes, indeed, it would be a lot easier," says Yoder, "and silicon-nitrogen polymers are known. The problem with these compounds in terms of biochemistry is that they're hydrolysable—water breaks them down very quickly—so it's not likely that you could have an aqueous biochemical system based on them. Now of course if you go to some other medium, such as liquid ammonia or liquid methane, then you might not have to worry so much about that."

That brings up the question of solvents: Is water the only possible solvent for biochemical reactions? One could not simply transfer terrestrial biochemistry wholesale into ammonia or alcohol or methane, of course, let alone into hydrochloric acid or mercury. At just 0.1 percent alcohol, we are legally drunk. At higher alcohol levels, cell membranes are destroyed and proteins and nucleic acids are denatured—they irreversibly lose their all-important three-dimensional structure. Other nonaqueous solvents are even more toxic. But is water preferable to other solvents in a general way, regardless of the details of life's chemical basis?

It's easy to make a list of the life-favoring properties of water. It is composed of two very abundant elements, and water itself is probably abundant in planetary systems, although not necessarily in liquid form. Compared with most alternative solvents, water has a high specific heat, meaning that its temperature remains relatively stable as heat is applied or withdrawn. This is an obvious advantage for a living system that can only operate within a limited range of temperatures. Vaporizing or freezing water also requires a relatively large application or withdrawal of heat, so water remains liquid longer than other solvents with similar boiling or freezing points. Even water's unusual property of expanding as it freezes seems to favor life: it causes ice to float, insulating the underlying water from further heat loss and thus slowing or preventing its freezing (recall the Antarctic lakes described in Chapter 2).

Water is a polar molecule: there is a partial separation of positive and negative charges across the molecule. The positively charged hydrogen atoms can interact with the negatively charged oxygen atoms of adjacent water molecules. These *hydrogen bonds* give water many of its unusual physical properties. Hydrogen bonds also form between water and charged regions within large organic molecules such as the DNA double helix, stabilizing their structure. In addition, nonpolar molecules, or portions of molecules, tend to bind with each other in the presence of water, thus reducing their overall contact with water. Such *hydrophobic* bonds between nonpolar amino acids play an important role in maintaining the integrity of many proteins. Phospholipids, which have both polar and nonpolar regions, readily coalesce in the presence of water to form bilayers—the basic structures of cell membranes.

All in all, water seems like the clear favorite as a medium for biochemical systems. Most probably, the first extraterrestrial Life we encounter will be water based. But still, we shouldn't count out other solvents, because these solvents have generally been studied much less than water, particularly in conditions of temperature and pressure remote from what we commonly encounter on Earth. The probable existence of a hydrocarbon ocean on Saturn's moon Titan (see Chapter 3) should at least remind us that supplies of alternative solvents are available, if only life can find a way to make use of them.

Not everyone shares the view that carbon and water offer the optimal chemical basis for life. Perhaps the most clearly stated opposing view came from the late Gerald Feinberg, a physicist at Columbia University, and Robert Shapiro, a chemist at New York University, who coauthored a 1980 book entitled *Life beyond Earth.* They dubbed carbon-and-water proponents "carbaquists" and suggested that carbaquists are people whose imagination has been stunted by their parochial experience of life on Earth.

Life, in Feinberg and Shapiro's definition, is the activity of a biosphere, and a biosphere is "a highly ordered system of matter and energy characterized by complex cycles that maintain or gradually increase the order of the system through the exchange of energy with the environment."[1] Many systems, they say, could have these properties. All the systems would need would be a source of, and sink for, free energy (thus permitting an energy flow), a medium suitable for the accumulation of order, and enough time for order to accumulate.

Where are these requirements satisfied? Sources and sinks for free energy exist almost everywhere, say Feinberg and Shapiro, so energy flows exist or can be created almost everywhere. Only in intergalactic space is it near impossible to set up an energy flow: not because there is no energy there—all space is permeated with microwave radiation left over from the Big Bang—but because there is no sink to which this energy can flow. Energy flows are especially rich at interfaces, such as the surface of the Earth, but they exist at many other locations too.

What media could use energy flows to accumulate order? To generate complexity, atoms need to react freely with each other, so chemical systems of life would likely be in fluid media—either liquids or dense gases. But the fluid would certainly not need to be water, and the complexity need not be built up in carbon compounds. In fact, say Feinberg and Shapiro, carbon may be a poor choice exactly because of its ability to form so many different compounds. These endless variants, most of which are not used by living organisms, act merely to hinder the useful biochemical processes. Elements like silicon or nitrogen may be more suitable for life because they are more selective in their associations.

Feinberg and Shapiro suggest that such alternative chemical life forms might already exist on or inside the Earth. They mention the possibility, for example, that organisms exist in lava flows ("lavobes") or in magma deep in the Earth ("magmobes"). These organisms would exploit thermal gradients or chemical energy sources within the molten rock, and complexity would build up in the rock's silicate chemistry—making use, for example, of the ability of minor elements like aluminum to interpolate themselves into the silicon-oxygen lattice in complex patterns.

A somewhat related proposal was put forward in the 1980s by the Scottish chemist Graham Cairns-Smith, who suggested that the crystalline silicate minerals in certain clays could carry complex information encoded in the pattern of irregularities in their crystal lattice.[2] This could happen most easily with minerals formed of booklike stacks of crystalline sheets, such as a class of silicate clays know as kaolinates. Each sheet, Cairns-Smith suggested, might "imprint" its pattern of irregularities onto the newly forming adjacent sheet, like ink spreading from one page of a book to the next.

Cairns-Smith put forward his system as a possible forerunner of organic life on Earth. He suggested that, on the early Earth, the surfaces of the clay crystals might have promoted the adsorption of organic compounds, which then underwent chemical reactions in complex spatiotemporal patterns, influenced by the underlying pattern of lattice irregularities. After the organic compounds developed their own complex, self-sustaining chemistry, however, the original crystalline matrix could have been abandoned.

The notion that clay crystals might have catalyzed a prebiotic surface chemistry is still quite popular today (see Chapter 1), but Cairns-Smith's central idea—the evolution of a heritable complexity in the clay crystals themselves—remains speculative. It is at least an intriguing thought-experiment for how life might translate itself gracefully from one medium (crystal structure) to another (organic chemistry). In general, switching media seems like a difficult task for mindless evolution to accomplish.

Could life exist in a nonchemical system? Feinberg and Shapiro suggest several habitats where purely physical life could exist. One is the interior of the Sun. The Sun consists of a plasma: a sea of positively charged atomic nuclei (mostly hydrogen and helium) and negatively charged electrons, all stirred up by intense magnetic fields. An energy flow is available as a radial (outward-directed) energy gradient within the Sun. Using this energy flow, the solar organisms ("plasmobes") would be formed by reciprocal interactions between the magnetic forces and the moving electrical charges, which might organize each other into ever more complex patterns. Feinberg and Shapiro see a parallel to the reciprocal interactions between nucleic acids and proteins in terrestrial life.

Another physical biosphere, according to Feinberg and Shapiro, could exist within or on the surface of neutron stars. The possibility of life on neutron stars has actually been discussed by a number of scientists, including Frank Drake.[3] As mentioned in Chapter 5, a neutron star is the tiny, superdense remnant of a supernova explosion. The core of the original star, no longer supported by the heat of nuclear reactions, collapses to a state in which gravitational compression overcomes the electrostatic repulsion between electrons and protons, which then fuse with each other, forming a sea of neutrons. Life might exist as patterns of bonded neutrons subject to the influence of the strong nuclear force.

From our point of view, the most striking feature of such a Life would be the breakneck speed of its metabolism; individual organisms might live and die within 10^{-15} seconds, and an entire "civilization" might rise and fall in a small fraction of a second. The speed of neutron-star processes also attracts Moravec. He has suggested that our robotic descendants might install a computer in a neutron star. Perhaps 10^{30} times more powerful than the human brain, the computer would be the "mind of last resort" for an entire robotic biosphere.

What about artificial life—life created intentionally? Intention has already played some role in the evolution of life on Earth, of course. Domesticated animals are an example. Even we humans are the product of our own intention to some degree: sexual selection, rewards and punishments, invasions and genocides—all have offered opportunities for the goal-oriented mind to intrude into the evolutionary process, for good or ill. But artificial life—especially when capitalized as Artificial Life and abbreviated as AL—refers primarily to a new field in which researchers attempt to create living or lifelike beings in computers.

"Living" or "lifelike"? That's the distinction between "strong" and "weak" AL. For a strong AL perspective, we talk by telephone with Tom Ray, a University of Oklahoma ecologist best known for his brainchild

Tierra, a computer within a computer (a "virtual computer") that houses a digital zoo of tiny, evolving organisms.[4]

A few years ago, at an early stage of his work, Ray offered a definition of life quite similar to the one given us by Jerry Joyce, the San Diego origin-of-life researcher. A system is living, Ray maintained, if it is self-replicating and capable of open-ended evolution. Since that time, he tells us, he has taken a somewhat more relaxed view. "Rather than making a list of requirements and saying, 'If it has everything on the list then it's alive,' the alternative is to make a long list of properties that are unique to living systems and to say, 'If it has any of these, then it has at least that property of life.'"

One reason for the change is that Ray recognizes that Lifes in different media may have different requirements. He rejects, for example, Stuart Kauffman's assertion that living things must earn a living in a thermodynamic sense. "That requirement may apply in a material medium, such as the world we live in. But in a digital medium the laws of thermodynamics simply don't exist. To say that you must obey the laws of thermodynamics to be alive simply excludes life in the digital medium. I take a broader view. I say, 'Here's a different medium, based on a completely different physics; and if we're going to admit the possibility of life there, we have to throw away our parochialisms based on the sample size of one that we have. We have to be willing to accept that it's not going to have the properties that we're familiar with.'"

In his writings, Ray has forcefully promoted his belief that Tierra is a real biosphere. "I Played God and Created Life in My Computer" was the title of a 1992 article of his that was translated into several languages. He also freely borrows terms from biology to describe what goes on inside Tierra, which he refers to as a "reserve for digital organisms."

The Tierran organisms are very simple: each is just a few dozen lines of computer code that allow it to propagate by copying itself to a new location in memory. More original, probably, is their home, a virtual parallel computer with a central processing unit (CPU) for each organism. The point of the virtual computer is threefold. First, it's a security measure: like the electrified fence surrounding Jurassic Park, it stops (or is intended to stop) the newly created beasts from invading the "real" world. ("In a real computer the creatures are just data," Ray tells us reassuringly.) Second, it allows for continuity: it prevents obsolescence as real computers change. Third, and most important, it partially overcomes the "brittleness" of the machine languages of real computers, a feature that has long impeded AL research.

"Brittleness" means the tendency of a computer to crash or generate error messages in response to even slight changes in the program. For example, a certain instruction, such as "CALL" or "GOTO," may tell the computer to move its attention to another, numerically labeled line of code ("GOTO 0043"). If line 0043 has for some reason become line 0053, the whole program may stop working. Thus, mutations are generally so damaging that evolution can't happen. In Ray's Tierran language, however, numerical instructions of this kind are eliminated. Instead, the software finds distant locations in the code by a template-matching procedure that Ray consciously borrowed from molecular biology: presented with a binary sequence such as 0110, for example, the computer searches in both directions along the code (invading the code of neighboring creatures, if need be) until it finds the complementary sequence, 1001, and executes the instruction at that location.

Thanks to innovations of this kind, the Tierran creatures are relatively robust in the face of mutations. And robust they need to be, because Ray has designed the code to mutate at a steady pace. In addition, a "Reaper" program kills off the digital creatures one by one as they begin to fill the virtual computer's memory. The Reaper selectively kills the older creatures and those whose codes generate a lot of error messages. Thus, there is a constant turnover of creatures, and rapid reproduction, rather than longevity, is the key to long-term survival.

Ray points out that, unlike many models of evolution such as Kauffman's, there is no explicit fitness function in Tierra: no formal measure of a creature's likelihood of surviving over the long term. It's simply the "survival of the survivors." This, Ray says, is a more truly Darwinian situation, in that the system sets up its own criteria for evolutionary fitness.

Ray has run Tierra over time spans corresponding to thousands of generations of the creatures within it. Evolution does indeed happen. Mutants appear that reproduce faster than the original creatures with which Ray seeded the system—for example, by copying more than one byte at each iteration. Complex ecologies build up within the Tierran biosphere. One mutant, for example, parasitizes the founder species: it uses the code of the founder species to reproduce itself, and thus can shorten its own code and reproduce faster. As the founder species gets outdone and begins to disappear from the biosphere, the mutant has a harder and harder time finding a creature to parasitize, so it in turn goes into a decline, and the founder species rebounds. This yo-yo behavior can continue indefinitely, as has been observed for some host-parasite pairs in terrestrial ecology. Host creatures may also develop immunity to parasites, or they may actually take over the CPU's of parasites that use their code ("hyperparasitism"). Cooperative arrangements can also appear, in which neighboring organisms, unable to reproduce by themselves, use parts of each others' code to reproduce successfully. These arrangements are in turn vulnerable to "cheaters," creatures that insinuate themselves between two cooperating creatures and take over the CPU's as they cross from one creature to the other. Thus, one message of Tierra seems to be a rather depressing one: that wherever you have evolution, you're going to have creatures ruthlessly exploiting one other.

Recently, with colleagues Kurt Thearling and Joseph Hart, Ray has used Tierra to explore the evolution of multicellular organisms.[5] "We consider this work to be an analogy to the study of processes that happened during the Cambrian explosion," Ray tells us. "What we're interested in is not the transition from the single-celled to the multicelled state, but the fantastically rich evolution that occurs after that transition has taken place."

To create a digital analog of this process, Ray's group seeded Tierra with "multithreaded" programs—programs that could be executed by multiple virtual CPUs, in the same way as the single human genome is executed by all the cells in our bodies. These creatures did indeed become multicellular, but contrary to their creators' hopes, the cells did not differentiate into specialized cell types or "tissues." Ray therefore went one step further and created founder cells that were specifically programmed to divide into two cell types—a "reproductive" type that contains the code for replication, and a "sensory" type that can acquire information about the computer environment.

Typically, seeding Tierra with these founder cells will lead to the development of ten-celled creatures, each containing eight sensory cells and two reproductive cells. No new cell types have emerged so far, however. "The only thing we've seen in that direction is the emergence of new genes in the genome," says Ray. "Some of the genes get duplicated during evolution, and each cell type expresses a different copy of the gene. Over time, the two copies mutate and diverge, so that they truly become different genes."

Ray has put a lot of effort into developing a network version of

Tierra, in which the Tierran creatures can move among linked computers. So far, he has the system running on a local area network, but he hopes ultimately to have thousands of versions of Tierra running on machines around the world, including those belonging to home personal-computer enthusiasts, all linked by the Internet. The idea is that Tierra would run in the background, or when the computers are not being used at all, such as during the night.

The point of Network Tierra is not only to increase the sheer size of the Tierra biosphere but also to introduce greater complexity into its environment. To prosper in the network, the creatures must develop some sense of where to go—how to home in on "good" computers that offer them plenty of computing time and that aren't turned off, for example. Even in the local network, Ray sees the beginning of social interactions. "There's a kind of flocking or mobbing behavior," he says. "All the creatures look around in the network, and they all gather data on the same machines and process it by the same algorithm, so they end up converging on a few 'best' machines. But the crowding makes those 'best' machines into the 'worst' machines. So an algorithm evolved to get around that problem: instead of choosing the 'best' machine, it avoided the 'worst' machine and remained neutral after that. This was a kind of social behavior that just evolved." In the planned global network, Ray expects to see other strategies appear: a tendency to remain on the dark side of the Earth, for example, where more computers stand idle.

Hearing of Ray's plans to involve the public in Tierra, and recalling Dan Werthimer's similar plans for SETI@home (Chapter 7), we can't help fantasizing about what might happen if the same computer were volunteered for both projects. In the fantasy, the evolving Tierran creatures detect and decipher an extraterrestrial signal in the SERENDIP data. Rather than alert the computer's owner, 15-year-old Albert Meaves of Ogallala, Nebraska, they manipulate him into facilitating an alien invasion of cyberspace. Now read on ...

All this, of course, presupposes a great deal of evolution from the level of complexity that Tierra has so far attained. But the thought motivates us to ask Ray how he personally relates to his digital creatures. Does he, for example, have a similar respect for them as for the organic creatures he has studied in the rain forests of Costa Rica? "Not yet, but the potentiality is there," he says. "The hope is that they will develop a rich enough structure that they will earn that respect; but right now, they're such simple little things, simpler than bacteria. I have a certain respect for bacteria, but I don't hesitate to kill them—and the same is true for digital organisms."

To Hans Moravec, what limits Tierra's evolution is sheer lack of computing power. "The reason Ray can't evolve anything other than simple predators is because it's such a small space—it's a microscopic drop of water. For intelligence to evolve—well, according to my calculations, just to hold one human intelligence takes a hundred trillion calculations a second, so to have a space where an intelligence could evolve would have to be much larger than that."

Just as Ray has had to deal with critics who see Tierra as "merely a simulation" of real life, so Moravec has had to confront people who say that computers and robots are not really conscious, but mimic consciousness. These kinds of disputes have venerable philosophical roots. In general, biological discoveries like William Harvey's discovery of the circulation of the blood and Darwin's theory of evolution fostered a gradually more positivist (antimetaphysical) attitude toward life, but a reaction set in with the work of the French philosopher Henri Bergson (1859–1941) who proposed the existence of an insubstantial "life force" (the *élan vital*) that imposed its will on matter and that presumably would not be present in life-imitating machines.

Through the middle of the twentieth century, however, ever-increasing knowledge about the functions of organs and tissues and the chemical basis of metabolism drove the vitalist point of view into the background again. Only one organ, the brain, retained a certain mystery. The special status of the brain was fostered by the lack of theories to explain how it worked, the lack of man-made analogs for it, the Cartesian tradition of mind-body dualism, and the widespread belief in the survival of consciousness after death. The notion of the brain's specialness is well illustrated in *Last and First Men*, the 1930 futurology epic by the British philosopher Olaf Stapledon.[6] At one point, Stapledon describes how, about 20 million years in the future, the Third Men engineer their successors, the Fourth Men, or "Great Brains":

> The dauntless experimenters succeeded at last in creating an organism which consisted of a brain twelve feet across, and a body most of which was reduced to a mere vestige upon the undersurface of the brain.... The fantastic organism was generated and matured in a building designed to house both it and the compli-

> cated machinery which was necessary to keep it alive. A self-regulating pump, electrically driven, served it as a heart. A chemical factory poured the necessary materials into its blood and removed waste products, thus taking the place of digestive organs and the usual battery of glands. Its lung consisted of a great room full of oxidizing tubes, through which a constant wind was driven by an electric fan. The same fan forced air through the artificial organs of speech. These organs were so constructed that the natural nerve-fibres, issuing from the speech centres of the brain, could stimulate appropriate electrical controls so as to produce sounds identical with those which they could have produced from a living throat and mouth.

Later versions of the Great Brains were induced to grow into a series of pigeonholes in a 40-foot-wide "brain turret" and were even implanted with Martian "telepathic units," but never in hundreds of millions of years of evolution did the process of thinking become dissociated from biological brain tissue. Yet only 5 years after *Last and First Men* was published, Stapledon's countryman, the 23-year-old Alan Turing, envisaged a universal computing machine, and the first "electronic brains" appeared a decade later. Real brains in vats suddenly seemed as quaint as seven-league boots.

Dualism—the notion that the mind can be separated from the operation of the brain or body—still has its adherents today, but the advent of computers and a succession of discoveries in neural science have greatly strengthened the materialist position. Thus, the interesting discussions tend to be, not about whether the mind can be explained in biological terms, but whether computers can be said to think in the same sense that humans do: When we operate a computer, does a mind function in the machine?

One person who has responded with an emphatic no to that question is John Searle, a philosopher at the University of California, Berkeley. To Searle, a computer is a syntactical engine—a device for generating strings of logically consistent symbols. It is not, however, a semantic engine: it does not recognize, utilize, or generate the meaning that may emerge from those syntactical structures. Minds do deal with meaning, so a computer is not a mind, and a mind is not a computer.[7]

To illustrate his point, Searle put forward his now-famous thought experiment of the "Chinese Room." A person who understands no

Chinese is put in a room with a list of instructions, written in English, for how to manipulate strings of Chinese characters. A Chinese speaker outside the room passes strings of Chinese characters under the door. The person inside the room produces new strings of Chinese characters, by reference to the list of instructions, and passes these strings out again. To the person outside, it seems that the person inside understands Chinese, but she doesn't. The person inside, Searle says, is analogous to a computer's CPU, and the instructions are analogous to a program. English is analogous to the computer's machine language, and Chinese to the language of humans dealing with the computer. In the same way as the person inside mechanically produces sensible Chinese discourse without understanding it, the computer mechanically produces sensible answers to questions posed to it without understanding their meaning.

The Chinese Room has become a favorite of introductory college philosophy courses, and numerous analyses of it, mostly critical of Searle's conclusion, can be found on the Internet, as well as in back numbers of the *New York Review of Books.* We expect Hans Moravec to have a critique ready to hand, and he doesn't disappoint us. "Understanding, which is Searle's main concern, is just an interpretation, an attribution that we make," he says. "It's not a substance. So it's perfectly reasonable for the Chinese-speaking person outside the room to attribute understanding based on the responses he receives from the room. And that's not at all contradicted by the mechanical operations done by the person inside the room. We're free to make all kinds of attributions, and Searle has constructed the example in such a way that we're most likely to make the mechanical attribution. The mistake is the idea that having the mechanical attribution invalidates the other attribution."

Moravec is fond of talking about the 1997 chess match in which the chess computer "Deep Blue" defeated the human world champion Garry Kasparov.[8] Besides being a landmark in artificial-intelligence research, the event illustrated how different people can justifiably make differing attributions to the same thing. "The Deep Blue team do not see intelligence in Deep Blue," Moravec says. "They see search functions, end-game tables, an opening book. They're too close to it, and they don't appreciate the chess it plays. Kasparov, on the other hand, knows how good it is. Several times during the match, he reported seeing signs of an alien intelligence in the machine."

Although Moravec admires the achievements of the Deep Blue

team, he thinks that highly specialized devices such as chess-playing computers are not the wave of the future. "A computer, given the right program, can do almost anything equally well," he says. "A computer only looks good or bad at certain tasks because we compare it to ourselves, but *we* are the ones that have the rugged landscape of abilities and disabilities. The computer is more like a water level, and that level is rising, so that there are some human skills that are well submerged already—a company wouldn't think of hiring clerks to do arithmetic, for example." Arithmetic is simply something evolution didn't prepare us for, Moravec contends, so our 100-million-MIPS brains spend minutes struggling to produce the solution to trivial problems like "What is 246 times 3078?"

Fifty years from now, Moravec predicts, computers will be much better than us in many areas and just equal to us in others. The last skills to be submerged—the Olympian peaks of the human mind—will include communicative skills such as negotiation and politics and artistic expression. "These are areas where you are really touching people at a deep psychological level—this was a life-or-death matter in our old tribal interactions, so its something we're really good at."

Of course, computers and robots won't stop improving at the point when they just match the human mind. Human robot-designing skills will become less and less relevant, as robots take an increasing role in designing and building their own successors. What then will be the fate of humans in a world where their talents are superfluous? In a 1988 book, *Mind Children*, Moravec predicted that humans would follow the robots' lead.[9] They would seek to improve their own intelligence and durability by augmenting their brains and bodies with artificial components, and eventually would transfer themselves entirely into computers—a feat that has since become known as "mind uploading."

More recently, however, Moravec has become concerned by the social problems that these activities would generate. "Somebody has a Napoleon complex," he says, "and simply mail-orders the army body and the supermind and starts lording it over the neighbors—well, that's just not going to fly. The neighbors will vote to prevent this, once the incidents start happening—maybe before. I think it will become like Switzerland, where you can't spit in the street and you have to be in bed by nine." (Moravec was born in Switzerland's underdog neighbor, Austria.)

Although Moravec has put forward a detailed description of the uploading technology—it involves a billion-fingered robot peeling away your brain, neuron by neuron, and rebuilding the equivalent circuitry in a computer, without interrupting your flow of consciousness—he now sees mind uploading as pretty irrelevant to humanity's future. "I don't see much point to it," he tell us. "Converting a human being into a robot is a lot like turning a horse cart into a race car—you end up with a tenth-rate race car, unless you totally destroy the original. Better to go back to the drawing board." The purpose-designed robots will be our descendants, and the best thing we will be able to do is bow out gracefully.

Whether humans remain as they are, upload themselves, or simply go extinct, the future lies with what Moravec calls "postbiological" beings. "Postorganic" would perhaps be a better descriptor, for these creatures, if they come into existence as Moravec predicts, will certainly attribute life and consciousness to themselves, even if they no longer have the "squishiness" required by Jerry Joyce's folk definition of life (Chapter 1). So they will still be biological entities—species in a biology of the future.

The robots (or whatever they are) will invade space in a big way, according to Moravec. They will virtually explode outward into the galaxy, led by an advance guard of macho engineer robots and followed by creatures of a more navel-contemplating disposition. Eventually, there will be little left for them to do except talk to each other and think.

This scenario, of course, brings Moravec up against the Fermi Paradox: If *our* robots are going to expand through the galaxy, why haven't *their* robots already reached Earth? Rather than take Ben Zuckerman's line and say that life on Earth is unique or atypical, Moravec accepts that other civilizations, some older than ours, may well be scattered through the galaxy. The reason we haven't seen their robotic ambassadors, he says, is that they're invisible. How so? In their eternal quest for more MIPS, Moravec says, the robots miniaturize their operations until they are working with matter on a scale much finer than we can see, much finer than the atomic scale, in fact. A wave of such reorganization can pass through a solar system without leaving its native inhabitants any the wiser.

Moravec laughs a lot as he leads us down the zanier byways of his imagination, but his broad, middle-European face maintains an earnest demeanor. We are probably among his more sympathetic listeners, within the scientific community at least. To us, it does seem hard to imagine any outcome to our present course of development other than the appearance of computers or robots smarter than ourselves. So may not our mission be the same as that of Cairns-Smith's clay crystals—to ignite a more brilliant flame of life, and then to peter out?

We do have reservations, though, about Moravec's timetable, which calls for human-level robots by the year 2040, full unemployment by 2050, and the end of organic intelligence a few decades later. Moravec himself has had to readjust his timetable a little: his original schedule called for entry-level domestic robots by 1998, but there is no sign of them as of this writing.

In general, roboticists have tended, by fair means or foul, to put an excessively positive spin on the state of their art. Around 1970, for example, researchers at Stanford made a film showing a robot purposefully manipulating blocks to create a ramp, up which it climbed to knock a small block off the top of the pile. In reality, the film was a montage of happy moments spliced together from an hours-long sequence of robotic blunderings. Another example: In his most recent book, *Robot*, Moravec shows a reconstruction of a room generated by a computer-vision system that he is developing for the robot Uranus. The scene shows chairs, floor, and walls in different colors. Although Moravec states quite explicitly that the objects have been hand colorized, the unwary reader will conclude that the robot has succeeded in distinguishing different objects in its surroundings, and perhaps even in identifying them. It hasn't; it has only mapped the layout of occupied and unoccupied space in the room—quite a feat in itself. Until the robot can distinguish and identify chairs and walls, and until it knows that chairs can be safely pushed aside, but walls not, most people will continue to do their own dusting.

A hundred years from now, or a hundred centuries from now—regardless of the time scale, it does seem plausible that postorganic beings will become the dominant life-forms on Earth. So such beings could perhaps exist right now on alien planets, just as many a science-fiction writer has postulated. We consider Moravec's answer to the Fermi Paradox (that aliens permeate us) a little too speculative: we prefer the original pro-SETI argument, namely, that interstellar travel is too difficult. The conclusion, then, is that we should be prepared for any extraterrestrial intelligence that we may contact to be of the postorganic variety. Maybe it will have lost all memory of its organic ancestors. Hearing it describe its robotic anatomy, we will say, "Put your creator on the line," but the response may be a puzzled, "Excuse me?"

Faced by all these real or potential Lifes—organic and inorganic, chemical and physical, natural and intentional—we of course come up against the fundamental question: What is life? This is a question that will be much easier to answer, we believe, when we have found and studied a good number of Lifes besides our own. But it's worth reviewing the issues that come into play.

On Earth, individual organisms are deeply embedded in the total biological and nonbiological environment. The limits of life are set by the totality of the environment, not by the organisms themselves. Fish multiply until they fill the pond, run out of smaller fish to eat, or attract too many fishermen. So a Life—a biosphere—is a *plenum*: it has no voids. When woodpeckers reach an oceanic island, they do not fill an empty niche: they *create* a niche by shouldering out a bunch of other creatures.

For this reason, we're sympathetic to Feinberg and Shapiro's definition of life, even if it is incomplete: Life is the operation of a biosphere. Organisms in isolation don't make too much sense. We could dream up a creature that violated most definitions of life: We could make a mule out of off-the-shelf chemicals, perhaps, so that it would neither have evolved nor self-organized, nor would it be capable of reproduction. We could fire this mule off into space with a lifetime supply of hay, so that it would be part of no ecosystem. It would continue to live, but it would be nothing but a futile splinter of a Life. It is the ensemble, the environment and the seething mass of creatures that fill it and interact with it, that gives life meaning.

The British chemist James Lovelock, along with Lynn Margulis, developed the "Gaia hypothesis"—the idea that Earth's organisms participate in feedback loops involving the atmosphere and oceans, and that these loops maintain the biosphere in a stable but nonequilibrium state over long periods of time.[10] We dislike the New-Agey overtones that have been added to the story—partly by Lovelock himself—in books with titles like *Gaia and God: An Ecofeminist Theology of Earth Healing.*[11] We would also point out that at least one event in Earth's history, the rise of atmospheric oxygen, could be seen as a total failure on Gaia's part—as a catastrophic pollution of the biosphere by selfish and self-destructive organisms. Even so, the Gaia hypothesis captures what is surely an important aspect of living systems—that organisms, through their energy exchanges with the inanimate environment, tend to maintain that environment in an out-of-equilibrium state. Indeed, when astronomers propose to find inhabited planets by detecting out-of-equilibrium gases in their atmospheres, as described in Chapter 5, it is Gaia that they are seeking.

Here on Earth, life is partitioned into individual organisms. We cannot be sure that this is a universal feature of biospheres. What this partitioning seems to allow is, first, the concentration of biomolecules to a level that permits metabolism and, second, natural selection and Darwinian evolution. But the very earliest life on Earth, at the point when organic chemistry became biochemistry, was probably not cellular. Concentration of biomolecules may have occurred through adsorption onto surfaces, drying, and other processes. How long a nonorganismal Life could persist, and what degree of complexity could be attained, is a difficult question that may eventually be resolved experimentally or even by observation of other biospheres; it seems unwise to prejudge the question by insisting that life be composed of discrete organisms.

Living organisms on Earth maintain a complex internal order by taking advantage of energy flows, or by creating such flows. This, of course, is how they exist in the face of the second law of thermodynamics. As Erwin Schrödinger pointed out, organisms excrete entropy, avoiding the "heat death" by accelerating the heat death of the rest of the universe.[12] According to Tom Ray, this principle is not a universal feature of life, because life could be instantiated in purely informational systems where the laws of thermodynamics are irrelevant. But any material Life that we encounter is surely likely to follow this rule.

Feinberg and Shapiro argued that many different chemical and physical systems could operate in this fashion. We agree in principle: There is absolutely no law that life must be built of carbon compounds dissolved in water. We would love to encounter Lifes in other media. But still, we remain "carbaquists" at heart, for four main reasons. First, carbon and water is the only proven recipe for life. Second, we have presented a story of the origins of carbon-based life that should be applicable widely in the universe, not just here on Earth. Third, many of the alternative Lifes that have been suggested, such as life in neutron stars, seem very difficult or impossible to detect. And fourth, many of the alternatives seem to offer much more limited opportunities for complexity than organic life does.

Some thinkers have emphasized the properties of organisms as informational systems. Theoretical physicist Lee Smolin, for example, has included in his definition of a living system the requirement that "its processes are governed by a program which is stored symbolically."[13] Certainly that is true of our own Life: it is hard to think of genomes as anything but stored programs. But autocatalytic sets of the kind envisaged by Kauffman contain no symbolically stored programs. How far life can develop without stored programs is uncertain, but it surely can at least get started.

The reason Smolin adds the "stored program" requirement is that he wants to exclude self-organized nonequilibrium systems, such as tornadoes or galaxies, that are not generally perceived as living. Others attempt to exclude such "low-level" nonequilibrium systems by adding a requirement that the system be chemical in nature, or that it evolve in a Darwinian fashion. The definition of life cited by Gerald Joyce (a "self-sustained chemical system capable of undergoing Darwinian evolution") adopts both those criteria. Like Smolin's requirement for a stored program, these two criteria are eminently satisfied in the case of our own Life, but it is not obvious that they should always be satisfied. In spite of our liking for organic life, we can't see any reason *in principle* that a Life should be based in chemical reactions rather than some other physical system, or any reason *in principle* that it should undergo Darwinian evolution rather than, say, modification by self-intention.

Because their view of life is so expansive, Feinberg and Shapiro are particularly at risk of having to include galaxies, tornadoes, candle flames, and the like in their cosmic bestiary. How does their definition deal with that problem? A biosphere, they say, is a "highly ordered system of matter and energy characterized by complex cycles that maintain or gradually increase the order of the system through the exchange of energy with the environment." Note the modifiers "highly" and "complex." These imply that there exist candidate systems whose applications for 'life' status have to be rejected because they are only "moderately" or "slightly" ordered, or are characterized only by "simple" cycles. What Feinberg and Shapiro never made clear is how one defines the level of order or complexity where 'life' status is warranted.

Feinberg and Shapiro's definition implies that life is continuous with non-life, which implies in turn that life is not what philosophers call a "natural kind"—it is not distinguished from non-life by any property open to empirical investigation. Philosopher Paul Churchland, of UCSD, offers a definition of life very similar to that of

Feinberg and Shapiro, but draws the conclusion of continuity more explicitly than they do. "The wiser lesson," he writes, "is that living systems are distinguished from nonliving systems only by degrees. There is no metaphysical gap to be bridged: only a smooth slope to be scaled, a slope measured in degrees of order and in degrees of self-regulation."[14]

The idea that life is not a natural kind also emerges from the "ingredient list" approach to a definition. Joyce's proffered definition is an example of an ingredient list, as is Tom Ray's original definition of life as "self-replicating and capable of open-ended evolution." Ingredient lists can get quite long. Ernst Mayr, for example, has said that it is not possible to offer a precise definition of living systems, but that they have many defining characteristics including the following: a complex adaptive organization, macromolecular structure, variability among individuals, genetic programs, "teleonomic" processes (see below), classes defined by common descent, unpredictability, and evolution by natural selection.[15]

Ingredient-list definitions imply continuity of life and non-life because, as with ingredient lists in cookery, other things can be made from the ingredients than the particular item that the list is intended to refer to. What an ingredient-list definition would need in order to make life distinct from non-life is a rationale for the combination of ingredients, but that is not usually supplied.

It will be recalled that Tom Ray revised his earlier definition of life. He said to us, "Rather than making a list of requirements and saying, 'If it has everything on the list then it's alive,' the alternative is to make a long list of properties that are unique to living systems and to say, 'If it has any of these, then it has at least that property of life.'" The problem with this approach, of course, is that it is questionable whether there *is* any single property that is unique to living systems, aside from Bergson's *élan vital*, which few if any thinkers still believe in. Thus, Ray's definition threatens to undermine the conceptual basis of Artificial Life: If life is not a natural kind, how can one say that a particular artificial construct does or does not have its single defining property, its combined list of properties, or even just one item out of the list?

We humans have a subjective experience of what it is like to be 'alive;' but this is really an experience of what it is like to have a human mind, not what it is like to be a carrot. It has been suggested that our intuitive sense that there is an *élan vital*, and that life is therefore a natural kind, is the product of on our intuitions about having a human mind.[16] Richard Cameron, a graduate student in philosophy at the University of Colorado who has prepared an illuminating survey of theories of the nature of life, calls this suggestion the "anthropomorphic thesis."[17] It seems to us that this thesis is at least partly correct. It may help explain why many writers, while offering definitions of life that are fundamentally discordant with the notion of life as a natural kind, nevertheless tend to downplay or ignore the discordance.

Because of our present state of ignorance, it is very difficult to decide whether life and non-life are continuous or not. Perhaps the strongest evidence for discontinuity is Stuart Kauffman's theoretical work, which suggests that autocatalytic sets suddenly "take off," once a certain threshold of complexity is reached. But we don't know if this is really what happened on the early Earth. If the takeoff phenomenon could be replicated in an artificial organic "soup," it would strengthen the argument that life and non-life are objectively distinct.

Even if life and non-life turn out not to be formally distinct, there may be rules of thumb that help us make the distinction in a practical context. For example, it may be appropriate to call a system 'living' if the imputation of purpose is helpful in understanding it. Here we echo the French molecular biologist Jacques Monod, who made "teleonomic behavior" (seemingly end-directed behavior, such as specific catalysis by a protein enzyme) a criterion for life.[18]

There are those who demand a definition of life as a guide for astrobiological research and exploration. The thought seems to be that astronauts should be provided with a checklist for use when probing squishy objects on alien planets. In truth, though, our current level of understanding can only suggest a range of possibilities. It may be wiser to be patient and let those squishy objects give us the answers.

SUGGESTED READING

Feinberg, G., and R. Shapiro. *Life beyond Earth: The Intelligent Earthling's Guide to Life in the Universe.* New York: Morrow, 1980.

Moravec, H. *Robot: Mere Machine to Transcendent Mind.* New York: Oxford University Press, 1998.

10

Many Worlds

Cosmology and the Anthropic Principle

"Chaotic inflation? I think that's really neat stuff!" says Ned Wright, his eyes sparkling. The UCLA astronomer is a connoisseur of cosmological theories, and this one—the brainchild of Russian-born physicist Andrei Linde—he savors especially. He thinks it could explain a lot about the universe we see around us, especially the curious ways in which it seems almost tailor-made to accommodate human beings.

But Wright is not one merely to admire ideas from afar; he likes to give them a thorough road test. "I try to make measurements that have the possibility of doing the greatest damage to the currently most popular theories," he tells us. And in the case of cosmological theories, that means aiming his telescope far beyond the planets and stars, beyond our entire galaxy and all the other visible galaxies, far back in space and time to the very backdrop of our cosmic scene, the darkly shining curtain that shrouds the Big Bang itself. Here, he believes, we will see writ large the secrets of the once-and-future cosmos.

We live, it seems, in a very strange universe. The more we try to understand it, the more it seems odd, cobbled together, contrived, as if someone had set a bunch of dials specifically to permit our existence.

"An obvious fix," Fred Hoyle called it.[1] So, was there a Fixer, or how did it come to be so hospitable to life?

The profoundest oddities about the universe are mathematical: certain coincidences and arbitrary-seeming constants that underlie the physics that we're familiar with. One example, though not the first, was pointed out by the English physicist Paul Dirac in 1937. Dirac noted that the ratio of the electrical and gravitational forces acting between an electron and a proton is roughly the same as the ratio of the size of the observable universe to the size of a proton: about 10^{39} in either case. Thinking that the coincidence must betoken some unknown connection between these phenomena, and realizing that the size of the observable universe is gradually increasing, Dirac suggested that the strength of the gravitational force must be gradually decreasing to keep the two ratios the same.

Recent high-precision measurements have shown that gravity is not in fact weakening, or at least not at the rate Dirac's hypothesis demanded. But long before these measurements were made, Princeton physicist Robert Dicke had laid out an "anthropic" argument for the large-number coincidence noted by Dirac—an argument that did not require any change in the strength of gravity over time.[2] Briefly put, Dicke argued that we could not have come into existence much earlier than we did because at least one generation of stars had to die and spew carbon into space before organic life could begin to evolve. The lifetime of a star, as Dicke showed, is controlled by the ratio of the electrical and gravitational forces. But the size of the observable universe is controlled by the time that has passed since the Big Bang, so the size of the universe *at the time we are able to observe it* is linked to the ratio of the electrical and gravitational forces. The "coincidence" is not a coincidence, in other words, if one keeps in mind our necessary position in time.

Besides coincidences between numbers, there are also quite a few other properties of the universe that seem to be suspiciously fine-tuned. One such property has to do with the crucial element carbon and how it is synthesized. No carbon was formed in the Big Bang; rather, as Fred Hoyle figured out in the 1950s, carbon is synthesized in stars by the fusion of beryllium and helium atoms. Hoyle realized, however, that in order for this reaction to proceed at any reasonable rate, carbon needed to have an appropriate "resonance," that is, an energy level that matched the combined energies of a beryllium and a helium atom plus the kinetic energy with which they collided. Only thus could a carbon atom be formed before the beryllium atom disintegrated. Motivated by Hoyle's prediction, his American colleague Willy Fowler confirmed the existence of this resonance experimentally. Without this arbitrary-seeming property, the universe could never have given rise to life as we know it.

Another example: The strength of the weak nuclear force, which mediates radioactive decay, has what seems to be an arbitrary value, but that value is necessary for carbon-based life to exist. If the value were slightly lower, supernovae would not explode, because the wave of neutrinos emerging from the collapsing star's core would not transfer sufficient energy to the star's outer layers. If the value were slightly higher, supernovae would again fail to explode, because the neutrinos would remain trapped in the star's core. Either way, without supernova explosions carbon would never reach interstellar space, and the cycle of life could not begin.[3]

There are perhaps a couple of dozen "facts" about the universe that have this flavor: They seem arbitrary, but they also seem absolutely necessary for life to have evolved. These facts include the values of several constants, the strength of certain forces, and the number of spatial dimensions.

One possible response to these apparently arbitrary (but life-friendly) facts is to question their arbitrariness. A scientist who does that with particular authority is Steven Weinberg of the University of Texas, Austin. Weinberg, along with Abdus Salam of Imperial College, London, played a key role in the development of "electroweak" theory, the theory that unifies the electromagnetic and weak nuclear forces. That theory has been so successful in predicting observations in particle physics (such as the discovery of the "w particle") that it has become known as the "standard model." The mass of the w particle is an example of a constant that might have seemed completely arbitrary if the particle had been discovered before Weinberg's work, but it is in fact predicted by the standard model.

"The standard model, in the form we developed it, does still have some constants in it that are unexplained," Weinberg tells us in a telephone conversation. "There are about eighteen free parameters in all—numbers that we just have to take from experiment." But Weinberg hopes that, as theories are developed that more completely unify the forces of nature, the arbitrariness of these numbers too will disappear. In fact, in collaboration with Howard Georgi and Helen Quinn, Weinberg has been able to reduce the number of free parameters by one. "We showed that if these forces are unified at some very high energy, you can predict the ratios of the three 'coupling constants,' so they are really only two independent constants."

With respect to Hoyle's carbon resonance, Weinberg disputes that there's anything very special about it. "I think that you'd expect a resonance with that energy anyway," he says. "It's not very sensitive to the fundamental constants of nature."

Another possible response to the coincidences is to say that life does not in fact have all the picky requirements that we usually attribute to it. We explored the possible existence of exotic forms of life in Chapter 9. There may not be anything so peculiar about the arbitrary constants if Feinberg and Shapiro were right about the ubiquity of life: the constants would still be arbitrary, for sure, but not arbitrary in any specific way that facilitates life and consciousness. Even if the constants had been such that the universe recollapsed on itself before a second of time had passed, maybe that single second would have been "deep time" to that universe's fast-living inhabitants! Of course, this argument doesn't invalidate the fact that the constants are fine-tuned for *our* existence, but that fact no longer has such great cosmic significance.

In spite of these possible responses, most cosmologists seem to believe that the universe is indeed surprisingly fine-tuned to permit life and consciousness to develop. To address this curious fact, many cosmologists rely in one way or another on what British astrophysicist Brandon Carter called the "anthropic principle," the idea that to understand the world around us we must take into account the selection effects imposed by our status as living observers. This principle often comes into tension with standard "Copernican" thinking, which asserts that there is nothing special about our position in, or view of, the cosmos.

One line of anthropic reasoning has been developed by physicist John Wheeler, who was until recently a colleague of Weinberg at the University of Texas, Austin. Wheeler was a student of quantum-mechanics pioneer Niels Bohr, and in 1939, he and Bohr published the first quantum-mechanical analysis of nuclear fission. Now in his eighties, Wheeler's stature as a physicist is such as to earn him the respectful attention of his peers, even if his current ideas seem a little exotic.

Wheeler believes that we in some sense participated in the creation of the universe we inhabit. His line of argument derives from observations in the strange world of quantum mechanics, the science that attempts to provide a quantitative description of the behavior of matter and radiation at the level of elementary particles. In quantum mechanics, certain properties of a particle are intrinsically unknowable. For example, one cannot hope to measure both an electron's momentum in a certain direction and its position in the same direction. A particle's behavior can be described only in terms of probabilities, and these probabilities are expressed mathematically as a particle's "wave function." It's thought that a photon, for example, travels as a distributed wave function along all its possible paths. When the photon is observed or measured (by a human observer or a photographic plate), the wave function "collapses" to a unique solution, corresponding to a particular path and a particular place where the photon actually ended up.

Such a system seems to allow for something akin to retrograde causation. Wheeler put forward "thought experiments" in which the apparent path taken by a photon was decided by the observer's choice to observe the photon in one way or another, even though the choice was made *after* the electron had completed most of its journey.[4] These thought experiments were later verified in the laboratory. Certainly, the apparent backward effect in time was a very brief one, measured in picoseconds. But then Wheeler showed the same effects on a cosmological scale. He made use of the phenomenon of gravitational lensing, in which light from a distant quasar takes two or more pathways around an intervening galaxy on its way to reaching us. By choosing how to observe the light waves that reach us via these alternative pathways (combining them by interference, or observing them separately), Wheeler seemingly affected the photons' properties when they set out on their journey, billions of years ago. From this it was but a short jump to the idea that we, as observers, influenced the properties of the Big Bang itself. We participated in the creation of the universe that we inhabit.

Ned Wright is not enamored of these ideas. He considers Wheeler's participatory theory to be "perhaps a little too much on the mystical side." He does acknowledge the possible force of Weinberg's argument: "It could be," he says, "that when we finally understand things like string theory all those ratios will be fixed once and for all. It could be that we'll be able to calculate the fine-structure constant [the "coupling constant" that governs the strength of the electromagnetic force in the interaction between light and elementary charged particles], just like we can now calculate pi. But there's other possibilities."

Those other possibilities include the idea that the universe we inhabit is not all there is—that there are other universes with different physical laws. If this is the case, anthropic reasoning provides a simple explanation for why the physical laws of *our* universe are such as to permit the appearance of life and consciousness: we simply have to find ourselves in such a universe, just as we have to find ourselves on clement Earth rather than on hellish Mercury. Chaotic inflation is one theory that seeks to account for the existence of many universes with diverse properties—a "multiverse" as it is sometimes called.

Chaotic inflation is an offshoot of the initial theories of inflation, which were put forward by Andrei Linde's teacher, Alexei Starobinsky, and by Alan Guth of M.I.T., around 1980. Theories of inflation were put forward to account for observed features of the universe that are not easily explained in classical models. For example, we do not infer any curvature of space from the density and motion of galaxies, yet general relativity implies that, in the overwhelming majority of possible cosmological models, space will be highly curved. Such curvature would prohibit the existence of life as we know it. So it would seem like an extremely improbable coincidence that we find ourselves in a flat universe.

The inflationary models eliminate the apparent improbability by proposing that the universe went through a period, ending about 10^{-35} seconds after the Big Bang, when space expanded exponentially, stretching itself by an enormous factor. This ensured that space would be essentially flat. Inflationary models also resolve other problems such as the apparent uniformity of the universe and the absence of magnetic monopoles, massive particles carrying a single magnetic charge that seemed to be required by classical models.[5]

Linde's "chaotic" or "self-reproducing" model explains why inflation starts in the first place, and why it stops. In this model, the scalar fields—fields that fill the universe and influence the properties of elementary particles—vary randomly in nature and strength at different locations. "It's like the temperature of a container of water," says Wright. "Some parts are cold, and the water looks like a solid; some parts are warm, and water looks like a liquid, and some parts are really hot, and water behaves like a gas. You have different physical behavior, depending on the value of the scalar field."

Because the values of the scalar fields vary randomly, there will be some places where the values are such as to permit inflation. These regions, therefore, inflate and come to dominate the universe. The entire system, then, is like a collection of bubbles that keep popping out of the scalar fields. We are in one of the bubbles, but there are others. "If there's only one way for it to happen," says Wright, "then all of the inflating bubbles are going to have the same laws of physics. But it could be that, when we understand string theory [a theory that proposes the existence of extra, "compactified" dimensions in addition to the four that we are familiar with], there will turn out to be four or five, or even several hundred ways you can get an inflating bubble, and in those bubbles you'll get different values of, for example, the fine structure constant." Thus, some of these bubbles (or universes) might offer habitats for life as we know it, some might offer habitats for life as we don't know it, and some might offer no habitats for life at all. But we will, of course, find ourselves in one of the life-friendly bubbles.

The main approach that Ned Wright is using to "road test" such cosmological theories is the observation of what is called the cosmic microwave background radiation. This radiation is a relic of the hot universe that emerged in the immediate aftermath of the Big Bang. For about the first 300,000 years, according to theory, photons should have wandered around like light in an opaque fog, constantly impeded by interactions with freely traveling electrons. In the process, the radiation should have equilibrated to a "blackbody" distribution, that is, to a distribution of wavelengths characteristic of a single temperature.

At the end of this period, the theory goes, the temperature of the universe fell to a point (about 3000°C) at which the electrons were able to unite with protons, forming hydrogen atoms. Once there were no longer any free electrons to interact with the photons, the latter traveled unimpeded through space. In essence, the universe became transparent. The radiation continued traveling through space ever since, but the expansion of the universe would have red-shifted, or "cooled," the relic photons to correspond to a temperature of around 3°K. Objects this cold radiate predominantly at millimeter wavelengths in the microwave region of the electromagnetic spectrum.

Arno Penzias and Robert Wilson, of Bell Labs, detected the microwave background radiation by accident in 1965. It appeared, at first, as a source of background noise in a simple horn antenna (now a National Historic Landmark). Try as they might, Penzias and Wilson couldn't eliminate the noise by any improvement in instrumental sensitivity. It took a not-yet-published paper by astrophysicists Jim Peebles, Robert Dicke, and David Wilkenson, of Princeton University, to explain the "noise." Working independently, and with impeccable timing, the Princeton group had predicted the properties of background radiation from the early universe. Penzias and Wilson soon realized that they were essentially looking back in time to the margin of that "fog bank" that was the early universe, and were seeing the light percolating out from the aftermath of the Big Bang that lay concealed in the fog-bank's heart. The Big Bang, of course, took place everywhere, but we can now see its image (or rather, that of the 300,000-year margin) only in places remote enough from us that it has taken the lifetime of the universe for the light to get from there to here. In other words, the cosmic background radiation emanates from the edge of the observable universe, beyond all the astronomical structures that we can study with telescopes.

Penzias and Wilson's discovery was a major piece of evidence that the universe did indeed begin in a Big Bang. But far more can be gleaned from the microwave background than that. For one thing, the microwave background should contain information about the smoothness, or lack or smoothness, in the density of matter at the 300,000-year "boundary." We know that, at very large scales, our present universe is quite smooth: a billion-light-year cube of space has about the same amount of matter in it as any other billion-light-year cube. But as one goes to smaller chunks of space, the universe becomes slightly "rough." The largest irregularities that we know about are densifications of matter like one dubbed the "Great Wall," a collection of clusters of galaxies about half a billion light-years across. At a finer scale, we have clusters of galaxies (about 5 million light-years across) and then galaxies (about 50,000 light-years across).

If the early universe had been perfectly smooth, such structures could never have arisen. The universe would have expanded, but evenly. So there would have been no galaxies, no stars, no planets, and no human beings. If very slight fluctuations in density had existed way back then, however, gravity would have exaggerated these fluctuations as the universe expanded: the slightly denser regions would have collapsed in on themselves, in much the same way as stars form from molecular clouds. Thus, the density differences would have increased, and we would get structures like the Great Wall, galaxy clusters, and galaxies.

If the density fluctuations in the early universe had been *too* great, however, the "roughness" of the present universe would be much more marked than we see. In particular, matter in large areas of space would have collapsed into "black holes," with little or nothing in between. (Black holes are regions where matter has collapsed gravitationally to a singular point of infinite density, creating a gravitational field so strong that nothing, not even radiation, can escape the black hole's "event horizon.") In fact, however, even the "supermassive" black holes that are believed to lie at the centers of some galaxies are only about the size of our solar system; most of the matter in the universe has not yet fallen into black holes, as far as we know. In other words, the amount of fluctuation in the early universe had to be just right, not too little and not too great, or we couldn't have been here.

Inflationary theories predict about the right amount of fluctuation in the early universe. These fluctuations were brought about by quantum-mechanical events that took place during the inflationary epoch—events that were then blown up into large-scale patches by the inflation itself. Depending on the time at which the individual quantum events occurred, they would have been subjected to inflation for differing periods of time and thus would have expanded into density fluctuations of a wide range of sizes. Since these fluctuations should have been present at the 300,000-year boundary, they should be detectable today in the microwave background radiation. That's because if a region at the 300,000-year boundary were slightly denser than its neighbor regions, radiation leaving that region would be red-shifted slightly more than light leaving the neighbor regions: it would lose energy as it climbed out of the "gravity well." In terms of the microwave background radiation, then, we would see a patch of sky that was slightly cooler than the surrounding regions. Furthermore, because of the timing issue just mentioned, inflationary theory predicts that there should be a wide range of patch sizes, with a peak at a patch size of about 1 degree of angle (which corresponds to about twice the apparent diameter of the Moon).

Ned Wright has been heavily involved in attempts to detect and measure these patches. It has been a major technical challenge, for two main reasons. First, the predicted amplitude of the fluctuations is very small—only about 1 part in 100,000. Second, measurements made from the surface of the Earth are hindered by the Earth's atmosphere, as well as by the Earth itself, which obstructs half the sky at any one time.

The first major success was achieved in 1990 by the Cosmic Microwave Background Explorer (COBE) satellite, which was designed and operated by a large group of scientists including Wright. (The project leader was John Mather of NASA's Goddard Space Flight Center.) The COBE results showed, first, that the microwave background does indeed have a distribution of wavelengths expected for blackbody radiation. In addition, however, the satellite was able to detect some of the slightly warmer and cooler patches that were predicted by inflationary theory (see Color Plate 14).

In spite of this success, the conclusions that could be drawn from the COBE data were limited, because COBE had rather coarse vision: it analyzed sectors of the sky several degrees across. "The spots we saw corresponded to regions of the universe that would now be about 5 billion light-years across," says Wright, "whereas even the largest structures in the universe that we have seen, such as the Great Wall, are only about 500 million light-years across, and clusters of galaxies are only about 5 million light-years across."

More-recent observations, made with ground-based and balloon-borne instruments, have extended the COBE data. In particular, the recent results are beginning to show a peak in the distribution of patch sizes at 1 degree of angle, as predicted by inflationary theory. There is far more information about the cosmos still hidden in the microwave background, however. To obtain this information, NASA and the European Space Agency are both planning missions to map the background at much greater resolution (to a small fraction of a degree) and with much greater sensitivity (to 1 part in 1,000,000). Wright is involved in planning NASA's mission, the Microwave Anisotropy Probe (MAP), which will be launched in 2000. The European Space Agency's mission, named Planck, will be launched about 5 years later.

Much of the still-hidden information bears on one of the most controversial problems in cosmology: The question of the manner in which the universe is expanding. As mentioned earlier, the universe is very close to "flat." That means its density is close to the "critical density": this is the density that allows the universe to expand at a rate just fast enough to avoid eventual recollapse. The fact that the universe is anywhere near the critical density now, billions of years after the Big Bang, means that it must have been extraordinarily near the critical density soon after the Big Bang. In fact, Wright tells us, if the density 1 nanosecond after the Big Bang had been just 1 part in 10^{23} greater than the critical density, the universe would by now already have recollapsed; while if it had been 1 part in 10^{23} less dense than the critical density, it would now be far more rarified than it is.

Inflationary theory predicts that the present universe is flat, and this prediction is supported by the observations already made on the microwave background. But astronomers have had a very hard time finding enough matter in the universe to bring it to the required critical density. In fact, the average density of all the matter that can be seen by astronomers, if it could be spread out evenly in space, would correspond to only about 0.1 atom per cubic meter. The critical density for the current universe is fifty times higher—about 5 atoms per cubic meter. So where is the missing mass?

At least part of the missing mass is accounted for by something called "dark matter." This is matter whose existence is inferred from its effect on the motions of galaxies and galaxy clusters, but which is not seen. The nature of the dark matter is highly contentious. Some astronomers think that it consists of ordinary baryonic (proton and neutron) matter. This could be clumped into brown-dwarf or planet-sized objects occupying a spherical halo around each galaxy. There have been attempts to detect these so-called MACHOS (massive compact halo objects) in the halo of our own galaxy by gravitational lensing: as a MACHO moves across the line of sight between us and a star in a neighboring galaxy, the light of that star should briefly intensify. Although several such lensing events have been detected, there has been some debate as to whether the lensing objects are really MACHOS in the halo of our galaxy, or whether they are ordinary stars in the neighboring galaxy.[6]

Another possibility is that the dark matter consists of WIMPS—weakly interacting massive particles. These hypothetical particles, with a mass at least a hundred times the mass of a proton, are called for in some supersymmetry models (theories that attempt to arrange particles in families according to their "spin"). Because the particles are uncharged, they would usually pass through a large object like the Earth without interacting with it. Still, experiments to detect their occasional interactions with ordinary matter are under way.

The dark matter associated with galaxies and galaxy clusters is still not enough to bring the density of the universe to the critical density needed by inflationary cosmologists; in fact, it only accounts for about 20 to 25 percent of the critical density. That then leaves about 75 percent unaccounted for: this seems to be neither ordinary matter nor dark matter.

Very recently, a possible new light on the missing mass has come from an unexpected source. Two groups have made observations of distant supernovae that suggest that the rate of expansion of the universe has been accelerating over its history. This is a very surprising result. If true, it suggests that there is some kind of negative pressure in the fabric of space-time that counteracts gravity. As the universe expands and objects move apart from each other, the gravitational attraction between objects weakens, so the negative pressure comes to dominate, making objects fly apart at an ever-increasing speed.

This negative pressure, which goes by the name of the cosmological constant, was originally proposed by Albert Einstein to explain why the universe, which he thought was static, did not collapse. He abandoned the idea after it was discovered that the universe is expanding (and, hence, counteracting gravity with the momentum of its expansion). If there is a cosmological constant after all, it would have an associated energy density and, hence (by $E = mc^2$), an equivalent mass. This mass could supply the missing 75 percent of the universe's mass needed to bring it to the critical density.

Steven Weinberg is fascinated by the cosmological constant, which, he says, is the most arbitrary-seeming of all the constants. "We can sort of guess how the other constants might arise," he tells us, "but the cosmological constant really seems mysterious." Weinberg suggests that the constant is actually made up of two terms, a term due to Einstein's field equations and a term due to the energy of the quantum vacuum. Both terms are extraordinarily large, but they very nearly cancel each other out. The constant that is observed is the difference between them.

"So why do they nearly cancel out?" he asks. In a recent attempt to answer the question, Weinberg (in collaboration with astronomers Hugo Martel and Paul Shapiro) invoked the very style of thinking that he has generally resisted—the anthropic principle.[7] "We made the assumption that the value of the constant is not fixed in advance but varies from one part of the universe to the other, maybe from one Big Bang to another. Then we tried to calculate what is the probability distribution of the values of the constant that would be reported by astronomers. Now of course that is very strongly affected by the fact that, in most cases, the values are such that you won't have any astronomers. Either the universe would be expanding too fast for galaxies to form, or it would recollapse so fast that there wouldn't be time for life to evolve. It's only in those cases where just by chance these terms nearly cancel that the universe lives long enough and expands slowly enough that there's a chance that life will form and that there will be astronomers. And then we asked, 'Taking this into account, what is the most likely value that astronomers would observe for the combined constant?' "

Weinberg and his colleagues found that the most likely value was such as to supply a large fraction of the critical density. "That's what now seems to be observed," he says. "It's premature to draw any definite conclusions, because the observations keep changing. But a consensus seems to be growing that a large fraction of the energy density needed to explain the expansion of the universe is in the form of the cosmological constant. This gives some encouragement to the anthropic principle."

Weinberg's mode of analysis obviously has some similarities to Linde's chaotic inflation model. But rather than being a physical model, it involves only a statistical analysis of probabilities. "Linde's work rests on a lot of assumptions about the underlying physics," Weinberg says, "including the existence of the scalar fields, and we just don't know enough yet to know if that's true. But it's a very interesting idea."

Ned Wright is inclined to doubt the existence of a cosmological constant, but if it doesn't exist, what could supply the missing mass? "There could be even more dark matter," he says. "It would have to behave in a fairly peculiar manner, though, because it doesn't show up in clusters of galaxies. That means it must be moving too fast to fall into them. And if it's moving that fast now, it was moving very fast in the beginning. And then instead of cold dark matter, it's hot dark matter. We know hot dark matter doesn't work as a model for structure formation, so it's a puzzle. If it's not a cosmological constant, it's got to be something else that's strange."

He tells us, however, that the existence of the constant can be verified or disproved by observation of the microwave background. As the photons of the microwave background travel through the universe toward us, they pass through regions of slightly greater-than-average or lower-than-average density. If a photon enters a region of greater-than-average density (a gravity well), it gains energy; and it loses the same amount of energy as it climbs out of the well—so the net effect is zero. But if there is a cosmological constant, the well becomes slightly shallower during the time the photon is crossing it; so the photon gives up less energy climbing out of the well than it gained entering the well, and it ends up with more energy than it started with. Conversely, a photon traveling across a gravity "hill" will end up with less energy than it started with. So a cosmological constant will reveal its presence by the presence of more-pronounced fluctuations in the microwave background (at certain large size scales) than would otherwise be seen. The MAP and Planck missions will be able to detect these effects, if they exist. "There might be something funny about those distant supernovae," Wright says, "but if two completely different techniques give the same answer, I'm going to sign on."

In spite of all the present uncertainties, Wright thinks that inflationary models are holding up very well. And Linde's chaotic model, with all its implications for why we inhabit such a life-friendly universe, is in the best shape of all. "It's such a simple idea," he says. "There's no fine-tuning involved, there's no carefully constructed potential function, no mysterious processes of entering and leaving inflation. You end up with some big, flat universes, and we're going to find ourselves in one of those."

Another "multiverse" model has been developed by Lee Smolin, a theoretical physicist at Pennsylvania State University.[8] This one involves black holes. In Smolin's model—perhaps "speculation" would be a fairer word—a black hole's singularity sprouts a new universe. That idea in itself is not new. Smolin's novel contribution, however, is to suggest that the laws of physics change slightly at the transition from the old universe to the new. This gives rise to a process resembling Darwinian evolution: If a newly sprouted universe has physical laws that, in comparison with those of its siblings, more strongly favor the rapid production of large numbers of black holes, then this universe will leave more descendants; so universes will gradually evolve to be better manufacturers of black holes. If the multiverse is infinitely old, of course, then the universes within it must have reached an equilibrium state, or cycle of states. Thus, we would imagine that our own universe would be nearly optimally designed to produce black holes—a presumption that is open to test. On the other hand, there is also the anthropic requirement for habitability, so perhaps our universe is actually a compromise or optimization of those two constraints. The two are related, however, since large stars are required both to produce black holes and to eject the carbon required for life.

Given the Darwinian flavor of Smolin's model, one could wonder whether the individual universes should be considered living organisms, and the entire multiverse a biosphere. If so, our own biosphere is reduced to a kind of endosymbiotic organelle—one of many, no doubt—within the larger organism. Smolin himself makes no claims of this kind, though others have done so on his behalf.[9]

In an influential 1983 paper, Brandon Carter used anthropic reasoning to make two rather surprising claims about life in the universe: First, extraterrestrial intelligence is likely to be extremely rare. And second, the human race is likely to go extinct before long.[10] Both arguments depend on a consideration of our position in time.

The first argument goes as follows. The total expected main-sequence lifetime of the Sun is about 10 billion years. The process of biological evolution on Earth, from the prebiotic soup to humans and technology, took roughly 4 billion years. These two numbers are very similar to each other, yet there's no obvious reason that they should be similar. After all, the processes of biological evolution don't seem to have any connection to the processes of nuclear burning in stars, so why shouldn't they differ by many orders of magnitude, rather than by a mere factor of 2 or thereabouts?

Because the Sun's lifetime and the evolution of technology are not connected, says Carter, it would seem an unreasonable coincidence to imagine that the *average* time required for the evolution of a technological civilization to evolve is similar to the time made available by the nuclear physics of stars. Much more likely is that we Earthlings are not average: either we took much *more* time to evolve than did most civilizations in the galaxy, or we took much *less* time. We are either the rearguard of the galactic caravan or its advance guard.

It's not likely, the argument goes on, that we are the rearguard. That's because there's no reason to expect us to have taken much longer than average to evolve than most civilizations—there's no obvious delaying factor here on Earth. (In addition, though Carter doesn't explicitly use it, one could invoke the Fermi Paradox here: If we are latecomers, why haven't we heard from our predecessors?) That then leaves the alternative possibility: that we have evolved much faster than the average. If we have evolved faster than average, we should expect, according to the bell-shaped curve of probabilities, to have evolved near to the latest possible date when external circumstances would permit it—which is to say, not long before the end of our Sun's existence. This is in fact when we did evolve. Thus, from an anthropic point of view, the "coincidence" between the time scales of our evolution and of nuclear burning in the Sun is totally unsurprising. And since most civilizations take much longer to evolve than we did, they in fact never do evolve because the death of their star terminates the evolutionary process before that can happen. So we are more or less alone in the galaxy.

Mario Livio, of the Space Telescope Science Institute in Baltimore, has recently offered a rebuttal to Carter's argument.[11] Livio suggests that a star's lifetime, and the time for intelligent life to evolve on a planet orbiting that star, may not be independent of each other, as Carter assumed. There is, after all, a relationship between a star's main-sequence lifetime and its luminosity: dimmer stars live longer. Conditions on planets orbiting dimmer stars may be such as to cause evolution toward intelligence to progress more slowly. As an example, Livio points out that an ozone shield, necessary for living creatures to emerge from the oceans onto the land, would develop more slowly on a planet orbiting a dim star, because such a star radiates less of the ultraviolet light required for the initial production of oxygen and ozone from water. Thus, there may be a general relationship such that, the dimmer the star, the longer intelligent life takes to appear.

As one goes from the brightest stars to increasingly dimmer classes of stars, however, the numbers of stars in each class increase. Because of these unequal numbers, a randomly chosen biosphere housing intelligent life is most likely to be one in which evolution took a long time—limited, of course, by the condition that this time be no longer than the lifetime of the star. Hence, says Livio, it is no coincidence that it took us a large fraction of the Sun's lifetime to evolve, and Carter's argument, based on that assumption, is cut off at the root.

In fact, Livio goes through further calculations, based on the rates of carbon production in stars, to suggest that the expected date for the first emergence of carbon-based intelligent civilizations should be about 10 billion years after the Big Bang, that is, about 3 billion years ago. Livio doesn't claim, however, to have shown that extraterrestrial intelligence is common, but only to have refuted one particular argument for its rarity.

The second of Carter's arguments—that we will soon go extinct—has been further elaborated and analyzed by physicist Richard Gott (who has developed his own "multiverse" model), as well as by the Canadian philosopher John Leslie.[12] It has come to be known as the "Doomsday Argument," and it runs something like this. Without evidence to the contrary, you do best to assume that you are a random selection from the set of all possible observers (that is a standard assumption in anthropic reasoning). You should therefore assume that your position in the series of all past, present, and future humans, ranked by birth date, is randomly selected. It is unlikely, then, that vastly more humans are still to be born than have been born already, because that would put you in an improbably tiny group of early-born individuals.

Now we know that the number of simultaneously living humans has increased greatly over time, so that the current (year 2000) world population actually equals about 10 percent of the cumulative total of all humans born up until now. In fact, it will only take one or two more centuries to reach such a large cumulative total that it will become apparent that you were a member of that "improbably early" contingent. If that is improbable, then it must conversely be probable that the human race will go extinct, or at least decrease greatly in numbers, within a couple centuries from now.

As we've described it, the Doomsday Argument sounds deceptively simple, but in fact it is extremely tricky. One brain-curdling analysis has been presented by Nick Bostrom, who is at the London School of Economics.[13] After fifty pages of thought experiments, calculation, and logical exegesis, Bostrom concludes that the Doomsday Argument has "so far withstood all attempts at refutation." We have nothing to add to his analysis, except this: Every human being, from the birth of our species until the recent past, could have gone through the same argument and reached the same conclusion, and on each occasion the prediction of imminent extinction would have turned out to be wrong. It doesn't seem too optimistic to hope that it will turn out wrong this time too.

The anthropic principle is obviously a mixed bag of ideas. In what Carter called its "weak" form, the principle simply reminds us not to take the Copernican view too far: not to forget that our observations are biased by a selection effect—by the necessity that we be here to observe. In fact, weak anthropic reasoning can lead back to a Copernican view. Faced with the "cosmic coincidences," for example, weak anthropic reasoning seems to call for an explanation in terms of some kind of multiverse, which is surely the ultimate in Copernican thinking. We're not just on one planet of many, orbiting one star among many, but we're also in one universe among many, albeit perhaps—in its capacity to support life—an uncommon universe.

The "strong" form of the anthropic principle is more explicitly teleological: It claims that the universe "had" to be such as to permit observers to exist in it. We can think of no good reason that the universe "had" to have such properties, unless the universe was created by ourselves (in some version of Wheeler's participatory model) or by an external Creator.

There are a few academics who do think that modern cosmology provides evidence for the existence of God. One of these is Walter Bradley, a professor of mechanical engineering at the University of Texas, Austin. A frequent speaker for the Campus Crusade for Christ, and a member of the ultraconservative web-based group "Leadership U," Bradley asserts that "one need never be ashamed of the intellectual respectability of belief in an intelligent creator; modern science has come down decisively on the side of the person who would posit such a belief."[14]

Weinberg groans audibly when we quote Bradley's remark. "I don't know his arguments, but I don't believe that," he says. "The more we learn about the universe, the more it seems to me that it is not governed by any principle in which morality or human life or love or justice, play any special role."

SUGGESTED READING

Barrow, J.D., and F.J. Tipler. *The Anthropic Cosmological Principle.* New York: Oxford University Press, 1986.

Rees, M. *Before the Beginning: Our Universe and Others.* Reading, MA: Perseus Books, 1997.

Weinberg, S. *Dreams of a Final Theory: The Scientist's Search for the Ultimate Laws of Nature.* New York: Pantheon Books, 1992.

Wheeler, J.A. *At Home in the Universe.* New York: Springer-Verlag, 1996.

Conclusions

Since the dawn of our species, an unquenchable curiosity has driven us to ask and re-ask the same questions: Who are we? Where do we come from? Are we alone? Myths offered the first answers: myths of creation and of spirit worlds. From here, one trail led to religion and to answers grounded in faith. But another trail led to philosophy—to reasoning. Pure reasoning at first, then reasoning aided by observation and experiment—what we now call science. To the first question, science has offered an astonishing answer: We are animals. To the second question, science has offered only a partial answer, but an equally astonishing one: We are descended from microbes. To the third question, which alone lends meaning to the other two, science has not yet offered any answer. Yet we are within reach of an answer—one that will surely be even more astonishing than the other two.

What brings an answer within our reach are two things. First, technological advances are allowing us to gather data that were totally unobtainable just a few years ago. Second, and perhaps more important, experts in many diverse disciplines have come together to forge a new science: astrobiology, exobiology, cosmic biology—call it what you will—it is a fundamentally new enterprise, a focus of intense excitement and energy, and a recipient of huge government resources. This science has just one ambition: To understand Life in its universal context and, in doing so, to understand ourselves.

We reject the point of view expressed to us by Stephen Gould, namely, that thinking about life in the universe is "fatuous speculation." Astrobiology, as an experimental science, must be guided by a discussion of what to look for and where to look for it. These questions, in turn, surely demand that we think about how life originates, where the building blocks of life come from, what range of habitats life can flourish in, how closely other Lifes should resemble our own, how evolution proceeds, how intelligent beings might make their presence known, and, not least, what life is.

As will be apparent from this book, the current climate in astrobiology resembles a brainstorming session, with many discordant voices going at it hammer and tongs, more than it does an orderly expression of consensus. Quite a bit of what is said will turn out to be wrong, but it still needs to be listened to.

Although no life has yet been found beyond Earth, enormous progress has been made in areas directly relevant to the search. One very productive area has been the study of the origins of stars and planetary systems. This research has offered two key insights. First, evolving stars do not suck up all the matter around them. Because of the initial random rotation of the clump of gas and dust from which stars are born, some material is commonly left in orbit around the star. Although the ultimate fate of this material has not been completely pinned down, a likely reason that the material ultimately disappears from view is that it gathers itself into planets. Thus, the study of starbirth has made it seem that the universe is rich in possible abodes for life—a view that has recently been validated by the actual detection of extrasolar planets.

The other insight that comes from the study of starbirth is the story of how organic compounds are formed in deep space and migrate into evolving solar systems, perhaps to be deposited on the surface of terrestrial planets. Admittedly, the story depends on a chain of observations and deductions, some strongly persuasive and some less so. The strongest part is probably the detection of the organic compounds in the molecular clouds where stars form. The study of these compounds by millimeter-wave and infrared spectroscopy is a thriving industry: new compounds are being added to the list faster than one can keep track of them. Less advanced, but a current area of interest, is the detection of organics in circumstellar disks. We feel confident that disks will turn out to be enormous reservoirs of prebiotic chemicals.

Complementing this search is the study of organic materials in comets, meteorites, and dust grains within our own solar system. All the signs are that there is a huge inventory of primordial organic material out there even now, 4.5 billion years after our solar system formed. The exact amounts and kinds of materials will become clearer from the further study of metereorites, from the study of grains collected by high-flying aircraft, and from sample-return missions such as the Stardust mission, which is scheduled to bring material emitted by comet "P/Wild," as well as interstellar dust grains, back to Earth in 2006.[1]

The most controversial part of this story is whether organic compounds reached the Earth's surface in amounts sufficient to build up a plausible prebiotic soup. The problem with answering this question is our uncertainty about the amount of infalling material and the composition and density of the Earth's early atmosphere, which determined what fraction of the material could escape incineration. The question of the atmosphere's composition also strongly affects the validity of alternative theories for the origin of prebiotic chemicals, especially Stanley Miller's theory that such chemicals were synthesized locally. It may be that we will eventually be able to get a handle on this question by studying the atmospheres of planets around young stars.

A second exciting area, and one that has received much well-justified publicity, has been the detection of extrasolar planets. So fast and furious are the discoveries coming, that the list of twenty or so planets will certainly be augmented by the time this book appears in print. Still, it's worth reemphasizing that, with the exception of Alex Wolszczan's pulsar planets, all the detected planets are likely to be gas giants. The detection of terrestrial planets must await even more ingenious technology, but the development of this technology, in the form of giant space-borne interferometers, is well under way. It seems very probable that by 2010 we will be describing the complete planetary systems around nearby stars, and that by 2020, we will be looking for signs of life in these planets' atmospheres.

A quite different area of research that also is yielding dramatic advances right now is molecular evolution. Again, this field is being driven by technological advances, this time in the development of fast, automatic gene-sequencing machines, as well as in the computer technology to assemble DNA-sequence information into entire genomes and to rapidly compare the genomes of different organisms. Just a few years ago, it seemed a pipe dream to completely sequence any organism. Now, we have complete genomes of a dozen or so microbes, a yeast, and a worm; and within 5 years or less, we will have the complete genomes of the fruit fly, the mouse, and *Homo sapiens.* In addition, the sequences of a few "test case" genes will be known for hundreds, or even thousands, of organisms.

What we know so far, thanks largely to the pioneering work of Carl Woese and his colleagues, is that all terrestrial organisms trace their ancestry back to a group of promiscuous "progenotes" that flourished in conditions of infernal heat. Most likely, these organisms earned a living by catalyzing electron transfers to and from minerals and gases in the environment. It remains unknown whether these organisms lived at deep-sea vents, in rocks far below the Earth's surface, or in the steaming pools of a planet still hot from its fiery birth.

To fully interpret the molecular data, two major tasks must be accomplished. First, the branch points in the molecular history need to be given dates: in spite of valiant efforts, the early stages of molecular evolution still lack any agreed-on time line. Second, connections need to be forged between molecular evolution and paleontology, including the microbial paleontology of Bill Schopf and others. These two tasks are closely related, and it seems likely that they will be accomplished within a few years. When they are, we will have a far more complete view of life's history on Earth than has been gained from either paleontology or molecular biology alone. The issues debated by Stephen Jay Gould and Simon Conway Morris—questions of the basic sense and strategy of the evolutionary process—should receive at least a partial answer.

Still, the extent to which the course of terrestrial evolution should have been mimicked on other planets is very uncertain, and will probably remain so until we are able to visit and explore them. Some principles seem to call out for universality. As Chris McKay puts it, life means that some creatures earn an honest living and other creatures eat them. That seems to be the lesson of terrestrial ecology and even of Tom Ray's Tierra. So this may be a bug-eat-bug universe. If so, the resulting pressures may commonly foster a trend toward complexification and the ability to react and learn. But whether that means that intelligence and technology is common in the galaxy, as the SETI proponents suggest, or whether these traits are as arbitrary and improbable as woodpecking, is unclear. We favor the former interpretation, and we find the opposing arguments (such as the Fermi Paradox) poorly conceived; but this point of view may be grounded as much in faith as in reason.

Whether SETI is destined to succeed or not, the technology of the search advances at the same frenetic pace as do the advances in electronics and computer science on which SETI feeds. While some developments are conceptual in nature, such as the recent interest in optical searches,[2] in addition to the traditional microwave search, the main story of SETI has been more sensitivity, more channels, more automation, and more telescopes. Given that SETI enjoys wide support within the astrobiology community, as well as among the general public, it seems likely that the US programs will eventually regain some degree of governmental support.

Although intelligent life is surely less common in the universe than nonintelligent life—perhaps orders of magnitude less common—it is also far more readily detectable. By 2020, we will be able to search for signs of nonintelligent life among a modest handful of nearby stars. But right now we have the ability to communicate with an intelligent civilization halfway across the galaxy, if it possesses merely the same level of technology as we do. Within that range lie hundreds of billions of stars. Such numbers can compensate for almost any plausible rarity of intelligence, provided only that life itself is common.

But is life common? And here we have to say that, even though some signs point in that direction, there is one major gap in our understanding: The quest to understand how the spark of life ignites from non-life is one area of research that has so far turned up more smoke than fire. Here, if anywhere, we need a God of the Gaps: a deus ex machina trundled in to get us past an awkward hole in the plot.

The main problem, of course, is that life as we know it depends on very long and specifically ordered polymers to catalyze reactions and to hold and transmit genetic information. The probability of even one such polymer assembling itself by chance is, as Fred Hoyle and others have emphasized, extremely small. One can therefore take the anthropic line and say: It happened once in the entire universe, and we have to be on the planet where it happened, so don't expect it to have happened again anywhere.

Given our ignorance of what did happen on the early Earth, we cannot definitively refute this argument, but we have a strong hunch that it is wrong. In particular, the rapid appearance of life—seemingly as soon as circumstances permitted it—speaks against that scenario. If the appearance of life from the prebiotic soup was highly unlikely, we would expect it to have been delayed as long as possible. We can't be sure how long the prebiotic soup would have persisted in the absence of life—a decrease in the rate of infalling material, combined with destruction of organics at the deep-sea vents, would have caused an eventual decline in the concentration of organics—but it certainly seems that life could have waited for several hundreds of millions of years at the least. The fact that life did not wait, but took advantage of pretty much the first opportunity offered, suggests that there is nothing improbable about that first spark.

The two scenarios advocated by Orgel and Kauffman—genelike polymers versus autocatalytic sets—are really just extremes of a continuum: the truth may be somewhere in the center. The first enzymes probably did a terrible job as far as specificity and speed are concerned, and the first genes probably did a terrible job as far as fidelity and durability are concerned. But it didn't really matter, as long as there was nothing better to compete with, and as long as there was plenty of energy to squander. What mattered was that the system as a whole had robustness, and that's where we find Kauffman's ideas a little more persuasive. We probably would barely recognize the first living system's enzymes as enzymes, or its genes as genes, and indeed these functions may have overlapped extensively.

Although the emergence of life on Earth is so problematic, there are avenues by which we can hope to find answers. The problem is tailor-made for laboratory experimentation. There is no reason that, within a reasonable time, we should not be able to create a plausible autocatalytic set and watch whether it does indeed "take on a life of its own." There is no reason that we should not create a wide range of potential RNA analogs or precursors to see whether any replicate themselves. And there is no reason that the RNA world should not be brought back to life, if it ever existed.

The question of the origin of terrestrial Life has a parallel in the question of the origin of our individual lives. Once conception was thought of as an infinitely mysterious event, requiring a hefty dose of divine intervention. Now we see it as just one of the remarkable, but basically explicable, processes that accompany the flow of life from one generation to the next. In due time, we may see the origin of terrestrial Life in the same way: as a predictable landmark in the flow of matter from organic chemistry into biochemistry. And the clincher, of course, will come when we have the opportunity to visit worlds where this process is happening right now.

The debate over life in the universe is driven, not just by the thousand particularities of the search, but by overarching issues that influence the thought patterns of scientists and laypeople alike. The tension between the Copernican viewpoint—the "principle of mediocrity," as Carl Sagan called it—and the anthropic style of reasoning is glaringly apparent. The Copernican viewpoint asserts that there is nothing special about our place in the universe. This attitude invites the conclusion that Earth-like planets are common, that they are commonly inhabited, and that some of the inhabitants are commonly intelligent. No wonder, then, that Sagan energetically promoted the search for extraterrestrial life and for signals from alien civilizations.

The limitations of the Copernican viewpoint have recently become clear. A hundred years ago, it was acceptable for scientists to believe that Mars was inhabited by intelligent creatures, even specifically by human beings. We now know that that is not the case: Mars houses subterranean microbes at best. And we also know that evolution, though it may show some trend toward complexity and intelligence, as Conway Morris believes, nevertheless is far too strongly influenced by contingency to turn out *Homo sapiens* twice in the universe, let alone on neighboring planets.

The anthropic (or anti-Copernican) viewpoint is that, in thinking about life in the universe, we should keep in mind the selection effect exerted by our own existence. We are bound to see life and an environment that can sustain life; we are bound to see a planet-girdled star. We should not generalize from the conditions we necessarily see to other parts of the universe where that necessity does not apply.

Some anthropic thinkers, such as Ben Zuckerman, seem to demand that we absolutely discount the knowledge whose acquisition depends on this selection effect. Earth has a large moon, he says, which stabilizes Earth's obliquity and makes life possible. But for anthropic reasons, we should forget about our moon. That leaves three other terrestrial planets—Mercury, Mars, and Venus—all of which lack large moons. Therefore, says Zuckerman, we should consider that we are batting zero for three, not one for four, in the moon game. This tally suggests that, in the universe at large, moons around terrestrial planets may well be close to nonexistent and, therefore, that life may be close to nonexistent too.

We believe that this argument takes the anthropic point of view too far for the following reason: Discovering that Earth has a large moon, we are entitled to think about how that moon came to be there. The leading theory is that it was created from the material propelled into orbit when a Mars-sized body struck the early Earth. We could, from theory and observation, develop some idea of how probable such a collision was. And we could observe accretion and debris disks around evolving stars to see if similar conditions exist there. Thus, without ever having found another example, we could form a reasonable hypothesis as to how common such moons may be, and the resulting number would likely be much greater than zero. This, then, is an intermediate stance between "We have a moon, so they should too" and "Our moon had to be there, so forget about it."

The same approach can be used to address the likelihood that a star will have planets. Until recently, we knew only that our Sun has planets. But our Sun had to have planets, so in the Zuckerman style of reasoning we were batting zero for zero in the planet game—not a good starting point for a planet search by any reckoning. Early models of planet formation, which involved interactions between two passing stars, did indeed call for planets to be vanishingly rare. The now-dominant model, however, which involves the collision of planetesimals within an evolving star's accretion disk, calls for planets to be commonplace. It was exactly the emergence of this theory from the earlier one that spurred the search for extrasolar planets, a search that has been so spectacularly rewarded.

Zuckerman is right that the anthropic principle bars us from playing a mindless numbers game. We should not claim that 25 percent of all terrestrial planets have large moons because that's what we see in our solar system, or that all stars have planets because our Sun has planets. But if the simplest model for how our circumstances came to be the way they are implies that these circumstances should be common, then we may justifiably extrapolate (with a greater or lesser degree of confidence) from our own circumstances to circumstances elsewhere. It is the combination of local observation and general understanding that has predictive power.

Even broader issues color the debate. Beliefs about extraterrestrial life—and the desirability or folly of looking for it—may in part be grounded in thoughts and feelings that have little logical connection to the subject. For example, between religion and the search for extraterrestrial life, there are sometimes striking parallels. This resemblance comes up particularly in the UFO field, as CSICOP chairman Paul Kurtz emphasized (Chapter 8). But there are hints of the same phenomenon even among scientists. Frank Drake, for example, has called the creatures we will contact "the immortals," who will teach us in turn the secret of immortality; and he has asserted his belief that contact will occur before the end of the twentieth century or during his own lifetime. Such remarks are uncannily similar to those of Jesus, when he said that "there be some of them that stand here, which shall not taste of death, till they have seen the kingdom of God come with power." Surely, regardless of the truth or otherwise of these statements, there are common underpinnings to the two men's utterances, separated though they are by 2000 years of cultural evolution. These underpinnings might include the fear of death, cosmic loneliness, the desire for some ultimate purpose to life, or an instinctive feel for the Divine. These traits have been very widely expressed throughout human history, but they don't seem to have much to do with the actual existence or nonexistence of life beyond Earth.

There is a second psychological thread that has to do with authority. Aliens are beloved of rebels, freethinkers, and libertarians, and are hated by control freaks. We have mentioned Lucretius, who recruited aliens for his assault on the ramparts of lofty Olympus. Lucretius has had many followers over the centuries. One was Tom Paine, the great voice of liberty on two continents. "To believe that God created a plurality of worlds," he wrote in 1793, "at least as numerous as what we call stars, renders the Christian system of faith at once little and ridiculous and scatters it in the mind like feathers in the air."[3] It was the organized, authoritarian aspects of Christianity, and its highly particularized doctrines, that irked Paine. He was a deist, not an atheist.

The reason that freethinkers love other worlds is that they are worlds beyond our control, and we beyond theirs. The universe is not a hierarchy, that is the message of Copernicus. The universe is not a Paris but a Los Angeles—hubless and endless. It promises the freedom of the megalopolis. But none of this tells us whether extraterrestrial life exists.

If we search and find no life—in our galaxy, say—the Copernican viewpoint will have to be modified, and profoundly so. We will still inhabit what seems like an ordinary planet circling an insignificant star, but we will be special for all that. Perhaps we will conclude that we were the result of a random but utterly improbable roll of the dice, and the rest of the universe is there just because there had to be that many rolls. Or maybe we'll conclude that we were indeed put here by design, and that whoever designed us—God, the engineers of a former world, or ourselves—had no choice but to build a universe, 15 billion light-years wide, to house us. Either way, the conclusion would be sobering, but it would not necessarily be bad. At the very least, it might encourage us to nurture a planet that was home to something so rare.

Our belief, though, is that the search will be rewarded, and soon. No argument is conclusive, and we may be influenced more than we realize by tribal yearnings. But the trail of discovery hangs heavy with the scent of life. What will we find? Who knows? Dragons surely not, but maybe something more outlandish. Humans surely not, but maybe someone wiser.

SUGGESTED READING

Harrison, A.A. *After Contact: The Human Response to Extraterrestrial Life*. New York: Plenum Press, 1997.

The Planetary Society. "Making Contact" [available at http://seti.planetary.org/Contact/default.html].

Notes

INTRODUCTION

1 Lucretius, *De Rerum Natura* (On nature), trans. R.M. Geer (Indianapolis: Bobbs-Merrill, 1965), bk. 2, line 1070.

CHAPTER 1

1 S.W. Hawking, *A Brief History of Time: From the Big Bang to Black Holes* (Toronto: Bantam Books, 1988), 115–16.
2 [http://www.icr.org/].
3 Lucretius, *De Rerum Natura*, bk. 2, line 1150.
4 A.I. Oparin, *Vozhiknovenie zhizny na aemle* (The origin of life) (New York: Dover, 1952).
5 S.L. Miller, "A Production of Amino Acids under Possible Primitive Earth Conditions," *Nature* 117 (1953): 528–29.
6 C. Sagan and C. Chyba, "The Early Faint Sun Paradox: Organic Shielding of Ultraviolet-Labile Greenhouse Gases," *Science* 276 (1997): 1217–21.
7 J.A. Brandes, et al., "Abiotic Nitrogen Reduction on the Early Earth," *Nature* 395 (1998): 365–67.
8 C. Chyba and S. Sagan, "Endogenous Production, Exogenous Delivery and Impact-Shock Synthesis of Organic Molecules: An Inventory for the Origins of Life," *Nature* 355 (1992): 125–32.
9 G. Wächtershäuser, "Evolution of the First Metabolic Cycles," *Proceedings of the National Academy of Sciences* 87 (1990): 200–204; G. Wächtershäuser,

"Before Enzymes and Templates: Theory of Surface Metabolism," *Microbiological Reviews* 52 (1998): 452–84.

10 A.D. Keefe, et al., "Investigation of the Prebiotic Synthesis of Amino Acids and RNA Bases from CO_2 Using FeS/H_2S as a Reducing Agent," *Proceedings of the National Academy of Sciences* 92 (1995): 11904–6.

11 J.W. Schopf, "Microfossils of the Early Archean Apex Chert: New Evidence of the Antiquity of Life," *Science* 260 (1993): 640–46.

12 M. Schidlowski, "A 3,800-Million-Year Isotopic Record of Life from Carbon in Sedimentary Rocks," *Nature* 333 (1988): 313–18.

13 S.J. Mojzsis, et al., "Evidence for Life on Earth before 3,800 Million Years Ago," *Nature* 384 (1996): 55–59.

14 L.E. Orgel, "The Origin of Life on the Earth," *Scientific American*, October 1994, 77–83. See also *The RNA World*, ed. R.F. Gesteland, T.R. Cech, and J.F. Atkins, 2nd ed. (Cold Spring Harbor: Cold Spring Harbor Laboratory Press, 1999).

15 J.P. Ferris, et al., "Synthesis of Long Prebiotic Oligomers on Mineral Surfaces," *Nature* 381 (1996): 59–61.

16 F. Hoyle and C. Wickramasinghe, *Diseases from Space* (London: Dent, 1979).

17 G.F. Joyce, "The RNA World: Before DNA and Protein," in *Extraterrestrials: Where Are They?*, eds. B. Zuckerman and M.H. Hart, 2nd ed. (Cambridge: Cambridge University Press, 1995), 139–51.

18 M.C. Wright, and G.F. Joyce, "Continuous in Vitro Evolution of Catalytic Function," *Science* 276 (1997): 614–17.

19 F. Hoyle, *The Intelligent Universe* (New York: Holt, Rinehart, and Winston, 1984), 19.

CHAPTER 2

1 C.P. McKay, "Relevance of Antarctic Microbial Ecosystems to Exobiology," in *Antarctic Microbiology*, ed. E.I. Friedmann (New York: Wiley-Liss, 1993), 593–601.

2 J.A. Nienow and E.I. Friedmann, "Terrestrial Lithophytic (Rock) Communities," in Friedmann, *Antarctic Microbiology*, 343–412.

3 C.P. McKay and E.I. Friedmann, "The Crytoendolithic Community in the Antarctic Cold Desert: Temperature Variations in Nature," *Polar Biology* 4 (1985): 19–25.

4 K.O. Stetter, "Hyperthermophilic Procaryotes," *FEMS Microbiology Reviews* 18 (1996): 149–59.

5 R. Jaenicke, "Glyceraldehyde-3-Phosphate Dehydrogenase from *Thermotoga maritima*: Strategies of Protein Stabilization," *FEMS Microbiology Reviews* 18 (1996): 215–24; D. W. Rice et al., "Insights into the Molecular Basis of Thermal Stability from the Structure Determination of *Pyrococcus furiosus* Glutamate Dehydrogenase," *FEMS Microbiology Reviews* 18 (1996): 105–17.

6 A. Kikuchi and K. Asai, "Reverse Gyrase: A Topoisomerase Which Introduces Positive Superhelical Turns into DNA," *Nature* 309 (1984): 677–81.

7 E.L. Shock, "Hydrothermal Systems as Environments for the Emergence of Life," in *Evolution of Hydrothermal Ecosystems on Earth (and Mars?)* (Ciba Foundation Symposium 202), ed. G.R. Bock and J.A. Goode (New York: Wiley, 1996), 40–60.

8 T.O. Stevens and J.P. McKinley, "Lithoautotrophic Microbial Ecosystems in

Deep Basalt Aquifers," *Science* 270 (1995): 450–54.

9 C.R. Woese, "Bacterial Evolution," *Microbiological Reviews* 51 (1987): 221–71.

10 R.F. Doolittle et al., "Determining Divergence Times of the Major Kingdoms of Living Organisms with a Protein Clock," *Science* 271 (1996): 470–77; M. Hasegawa and W. Fitch, "Dating the Cenancestor of Organisms," *Science* 274 (1996): 1750; J.P. Gogarten et al., "Dating the Cenancestor of Organisms," *Science* 274 (1996): 1750–51.

11 P. Forterre et al., "The Unique DNA Topology and DNA Topoisomerases of Hyperthermophilic Archaea," *FEMS Microbiology Reviews* 18 (1996): 237–48.

12 C. Woese "The Universal Ancestor," *Proceedings of the National Academy of Sciences* 95 (1998): 6854–59.

13 L. Margulis and R. Fester, eds., *Symbiosis as a Source of Evolutionary Innovation: Speciation and Morphogenesis* (Cambridge: MIT Press, 1991); M.W. Gray, G. Burger, and B.F. Lang, "Mitochondrial Evolution," *Science* 283 (1991): 1476–1481.

14 T. Hashimoto, et al., "Secondary Absence of Mitochondria in *Giardia lamblia* and *Trichomonas vaginalis* Revealed by Valyl-tRNA Synthetase Phylogeny," *Proceedings of the National Academy of Sciences* 95 (1998): 6860–65.

15 W. Martin and M. Müller, "The Hydrogen Hypothesis for the First Eukaryote," *Nature* 392 (1998): 37–41.

CHAPTER 3

1 S.J. Dick, *The Biological Universe: The Twentieth-Century Extraterrestrial Life Debate and the Limits of Science* (Cambridge: Cambridge University Press, 1996), 74–75.

2 Ibid. 99.

3 H.G. Wells, *The War of the Worlds* (1897; reprint, Buccaneer Books, 1983).

4 K. Lasswitz, *Auf Zwei Planeten* (Two planets), H.R. Rudnick, trans. (1897; reprint, Carbondale: Southern Illinois University Press, 1971).

5 S. Sagan and P. Fox, "The Canals of Mars: An Assessment after Mariner 9," *Icarus* 25 (1975): 602–12.

6 C.P. McKay, "Looking for Life on Mars," *Astronomy*, August 1997, 38–43.

7 M.H. Carr, *Water on Mars* (New York: Oxford University Press, 1996).

8 V.R. Baker, et al., "Ancient Oceans, Ice Sheets and the Hydrological Cycle on Mars," *Nature* 352 (1991): 589–94; J.S. Kargel and R.G. Strom, "Global Climatic Change on Mars," *Scientific American*, November 1996, 80–88.

9 T.J. Parker, et al., "Coastal Geomorphology of the Martian Northern Plains," *Journal of Geophysical Research* 98 (1993): 11061–78.

10 F. Forget and R.T. Pierrehumbert, "Warming Early Mars with Carbon Dioxide Clouds That Scatter Infrared Radiation," *Science* 278 (1997): 1273–76.

11 D.S. McKay, et al., "Search for Past Life on Mars: Possible Relic Biogenic Activity in Martian Meteorite ALH84001," *Science* 273 (1996): 924–30; E.K. Gibson, Jr., et al., "The Case for Relic Life on Mars," *Scientific American*, December 1997, 58–65. For summaries and critiques of follow-up studies, see A. Treiman, "Recent Scientific Papers on ALH84001 Explained, with Insightful and Totally Objective Commentaries." [available at http://cass.jsc.nasa.gov/lpi/meteorites/alhnpap.html].

12 B. Gladman, "Destination Earth: Martian Meteorite Delivery," *Icarus* 130 (1997): 228–46.

13 J.L. Bada, et al., "A Search for Endogenous Amino Acids in Martian Meteorite ALH84001," *Science* 279 (1998): 362–65.
14 A.J.T. Jull, et al., "Isotopic Evidence for a Terrestrial Source of Organic Compounds Found in Martian Meteorites Allan Hills 84001 and Elephant Moraine 79001," *Science* 279 (1998): 366–69.
15 J.L. Kirschvink, A.T. Maine, and H. Vali, "Paleomagnetic Evidence of a Low-Temperature Origin of Carbonate in the Martian Meteorite ALH84001," *Science* 275 (1977): 1629–33.
16 J.P. Bradley, H.Y. McSween, Jr., and R.P. Harvey, "Epitaxial Growth of Nanosphere Magnetite in Martian Meteorite ALH84001: Implication for Biogenic Mineralization," *Meteoritic and Planetary Science* 33 (1998): 765–73.
17 E.I. Friedmann, J. Wierzchos, and C. Ascaso, "Chains of Magnetite Crystals in Allan Hills 84001: Evidence of Biological Origin," Lunar and Planetary Institute Workshop on Martian Meteorite ALH 84001." [available at http://cass.jsc.nasa.gov/lpi/meteorites/marsmet98.html].
18 J.P. Bradley, R.P. Harvey, and H.Y. McSween, Jr., "No 'Nanofossils' in Martian Meteorite Orthopyroxenite," *Nature* 390 (1997): 454. The letter is followed by a reply from McKay's group.
19 R.A. Kerr, "Requiem for Life on Mars? Support for Microbes Fades," *Science* 282 (1998): 1398–1400.
20 D.W.G. Sears and T.A. Kral, "Martian 'Microfossils' in Lunar Meteorites?" *Meteoritic and Planetary Science* 33 (1988): 791–94.
21 J.A. Barrat et al., "Formation of Carbonates in the Tatahouine Meteorite," *Science* 280 (1998): 412–14.
22 E.I. Friedmann, "The Antarctic Cold Desert and the Search for Traces of Life on Mars," *Advances in Space Research* 6 (1986): 265–68.
23 C.P. McKay, O.B. Toon, and J.F. Kasting, "Making Mars Habitable,"*Nature* 352 (1991): 489–96.
24 E.I. Friedmann, "Extreme Environments and Exobiology," *Giornale Botanico Italiano* 127 (1993): 369–76; E.I. Friedmann, M. Hua, and R. Ocampo-Friedmann, "Terraforming Mars: Dissolution of Carbonate Rocks by Cyanobacteria," *Journal of the British Interplanetary Society* 46 (1993): 291–92.
25 M.J. Fogg, "Terraforming: A Review for Environmentalists," *The Environmentalist* 13 (1993): 7–17.
26 M. Hart, "Habitable Zones around Main Sequence Stars," *Icarus* 37 (1979): 351–57.
27 J.F. Kasting, D.P. Whitmire, and R.T. Reynolds, "Habitable Zones around Main Sequence Stars," *Icarus* 101 (1993): 108–28.
28 M.A. Slade, B.J. Butler, and D.O. Muhleman, "Mercury Radar Imaging: Evidence for Polar Ice," *Science* 258 (1992): 635–40. See also the NASA site, "Goldstone/VLA Maps of Mercury," [available at http://wireless.jpl.nasa.gov/RADAR/mercvla.html].
29 W.C. Feldman et al., "Fluxes of Fast and Epithermal Neutrons from Lunar Prospector: Evidence for Water Ice at the Lunar Poles," *Science* 281 (1998): 1496–1500. See also the NASA site, "Ice on the Moon," [available at http://nssdc.gsfc.nasa.gov/planetary/ice/ice_moon.html].
30 See the Galileo mission's home page [available at http://www.jpl.nasa.gov/galileo/index.html].
31 K.K. Khurana et al., "Induced Magnetic Fields as Evidence for Subsurface Oceans in Europa and Callisto," *Nature* 395 (1998): 777–80.
32 C. Sagan, W.R. Thompson, and B.N. Khare, "Titan: A Laboratory for

Prebiological Organic Chemistry," *Accounts of Chemical Research* 25 (1992): 286–92; C. Sagan, "The Search for Extraterrestrial Life," *Scientific American*, October 1994, 93–99.

33 See the Cassini mission's home page [available at http//www.jpl.nasa.gov/cassini/]. See also the Huygens probe home page [available at http://sci.esa.int/huygens/].

CHAPTER 4

1 R.P. Kirschner, "The Earth's Elements," *Scientific American*, October 1994, 59–65. See also D. Arnett, and G. Bazan, "Nucleosynthesis in Stars: Recent Developments," *Science* 276 (1997): 1359–62.

2 D. Johnstone, "SCUBA Scan-Mapping with the JCMT" [available at http://www.cita.utoronto.ca/~johnston/jcmt.html#letter]; R. Chini, et al., "Dust Filaments and Star Formation in OMC-2 and OMC-3," *Astrophysical Journal* 474 (1997): L135–38.

3 F.H. Shu, "Self-Similar Collapse of Isothermal Spheres and Star Formation," *Astrophysical Journal* 214 (1977): 488–97.

4 S. Terebey, F.H. Shu, and P. Cassen, "The Collapse of the Cores of Slowly Rotating Isothermal Clouds," *Astrophysical Journal* 286 (1984): 529–51.

5 C.J. Burrows, et al., "Hubble Space Telescope Observations of the Disk and Jet of HH 30," *Astrophysical Journal* 473 (1996): 437–51.

6 C.R. O'Dell and S.V.W. Beckwith, "Young Stars and Their Surroundings," *Science* 276 (1997): 1355–58. For more Hubble telescope images of Orion, see "CITA Orion Nebula Research" [available at http://www.cita.utoronto.ca/~johnston/orion.html#figures].

7 A.I. Sargent and S.V.W. Beckwith, "Kinematics of the Circumstellar Gas of HL Tauri and R Monocerotis," *Astrophysical Journal* 323 (1987): 294–305.

8 J.I. Lunine, "Physics and Chemistry of the Solar Nebula," *Origins of Life and the Evolution of the Biosphere* 27 (1997): 205–24.

9 [http://stardust.jpl.nasa.gov/comets/giotto.html].

10 J.X. Luu and D.C. Jewitt, "The Kuiper belt," *Scientific American*, May 1996, 46–52.

11 [http://www.jpl.nasa.gov/sl9/].

12 D.W. Koerner, et al., "Mid-Infrared Imaging of a Circumstellar Disk around HR 4796: Mapping the Debris of Planetary Formation," *Astrophysical Journal* 503 (1998): L83–87.

13 R. Jayawardhana, et al., "A Dust Disk Surrounding the Young A star HR 4796A," *Astrophysical Journal* 503 (1998): L79–82.

14 C.J. Burrows, et al., "HST Observations of the Beta Pictoris Circumstellar Disk," Bulletin of the American Astronomical Society, no. 187, abstract 32.05 (1995).

CHAPTER 5

1 G. Gatewood, "Lalande 21185," Bulletin of the American Astronomical Society, no. 28, 885 (1996).

2 A. Wolszczan, and D.A. Frail, "A Planetary System around the Millisecond Pulsar PSR 1257 + 12," *Nature* 355 (1992): 145–47.

3 A. Wolszczan, "Confirmation of Earth-Mass Planets Orbiting the Millisecond Pulsar PSR B1257 + 12," *Science* 264 (1994): 538–42.
4 K. Scherer, et al., "A Pulsar, the Heliosphere, and Pioneer 10: Probable Mimicking of a Planet of PSR B1257 + 12 by Solar Rotation," *Science* 278 (1997): 1919–21.
5 A. Wolszczan, "Further Observations of the Planets Pulsar" [available at http://www.stsci.edu/stsci/meetings/planets/abs/prog31.html].
6 K.J. Joshi and F.A. Rasio, "Distant Companions and Planets around Millisecond Pulsars" [available at http://xxx.lpthe.jussieu.fr/abs/astro-ph/9610213].
7 M. Mayor and D. Queloz, "A Jupiter-Mass Companion to a Solar-type Star," *Nature* 378 (1995): 355–59.
8 D.M. Williams, J.F. Kasting, and R.A. Wade, "Habitable Moons around Extrasolar Giant Planets," *Nature* 385 (1997): 234–36.
9 C. Dominik, et al., "A Vega-like Disk Associated with the Planetary System of ρ^1 Cancri," *Astronomy and Astrophysics* 329 (1998): L53–56.
10 D.E. Trilling, and R.H. Brown, "A Circumstellar Dust Disk around a Star with a Known Planetary Companion," *Nature* 395 (1998): 775–77.
11 J. Greaves, et al., "A Dust Ring Around Epsilon Eridani: Analog to the Young Solar System," *Astrophysical Journal* 511 (1998): L133–L137.
12 Kepler Mission [available at: http://www.kepler.arc.nasa.gov].
13 Palomar Testbed Interferometer [available at: http://huey.jpl.nasa.gov/palomar/].
14 Keck Interferometer [available at: http://huey.jpl.nasa.gov/keck/].
15 Space Interferometry Mission [available at: http://sim.jpl.nasa.gov/].
16 R. Bracewell, "Detecting Nonsolar Planets by Spinning Infrared Interferometer," *Nature* 274 (1978): 780–81.

CHAPTER 6

1 G. Johnson, *Fire in the Mind: Science, Faith, and the Search for Order* (New York: Knopf, 1985).
2 P. Bak, C. Tang, and K. Wiesenfeld, "Self-Organized Criticality: An Explanation of 1/f Noise," *Physical Review Letters* 59 (1987): 381–84; P. Bak, and K. Chen, "Self-organized criticality," *Scientific American*, January 1991, 46–54.
3 For a review, see A. Mehta, and G.C. Barker, "The Dynamics of Sand," *Reports on Progress in Physics* 57 (1994): 383–416.
4 S.A. Kauffman, *The Origins of Order: Self-Organization and Selection in Evolution.* (New York: Oxford University Press, 1993).
5 C. Huber, and G. Wächtershäuser, "Peptides by Activation of Amino Acids with CO on (Ni,Fe)S Surfaces: Implications for the Origin of Life," *Science* 281 (1998): 670–72.
6 S.J. Gould, "The Evolution of Life on the Earth," *Scientific American*, October 1994, 85–91.
7 N. Eldredge and S. Gould, "Punctuated Equilibria: An Alternative to Phyletic Gradualism," in *Models of Paleobiology*, ed. T.J.M. Schopf (San Francisco: Freeman, Cooper, 1972), 82–115.
8 H.B. Whittington, *The Burgess Shale* (New Haven: Yale University Press, 1985); D.E.G. Briggs, D.H. Erwin, and F.J. Collier, *The Fossils of the Burgess Shale*

(Washington: Smithsonian Institution Press, 1994).

9 L. Ramsköld and H. Xianguang, "New Early Cambrian Animal and Onychophoran Affinities of Enigmatic Metazoans," *Nature* 351 (1991): 225–28.

10 S.B. Carroll, "Homeotic Genes and the Evolution of Arthropods and Chordates," *Nature* 376 (1995): 479–85; R.A. Raff, *The Shape of Life: Genes, Development, and the Evolution of Animal Form* (Chicago: University of Chicago Press, 1996).

11 M. Eigen, and R. Winkler Oswatitsch, *Steps towards Life: A Perspective on Evolution* (New York: Oxford University Press, 1992).

12 B. Hayes, "The Invention of the Genetic Code," *American Scientist* (January/February 1998): 9–14.

13 D. Haig and L.D. Hurst, "A Quantitative Measure of Error Minimization in the Genetic Code," *Journal of Molecular Evolution* 33 (1991): 412–417. S.J. Freeland and L.D. Hurst, "The genetic code is one in a million," *Journal of Molecular Evolution* 47 (1998) 238–248.

14 R.D. Knight and L.F. Landweber, "Rhyme or Reason: RNA-Arginine Interactions and the Genetic Code," *Chemistry and Biology* 5 (1998): R215–20.

15 E. Mayr, "The Search for Extraterrestrial Intelligence," in Zuckerman and Hart, *Extraterrestrials: Where Are They?* 152–56; E. Mayr, "Can SETI Succeed? Not Likely" [available at: http://seti.planetary.org/Contact/debate/debate-no.html].

16 J. Diamond, "Alone in a Crowded Universe," in Zuckerman and Hart, *Extraterrestrials: Where Are They?* 157–64.

17 W.H. Calvin, "The Emergence of Intelligence," *Scientific American*, October 1994, 101–7.

CHAPTER 7

1 Dick, *Biological Universe*, chap. 8.

2 G. Cocconi and P. Morrison, "Searching for Interstellar Communications," *Nature* 184 (1959):844.

3 B.M. Oliver and J. Billingham, *Project Cyclops: A Design Study of a System for Detecting Extraterrestrial Intelligent Life* (Moffet Field, CA: NASA Ames Research Center, 1971).

4 [http://www.seti.org/phoenix/].

5 [http://seti.planetary.org/SERENDIP/default.html].

6 [http://mc.harvard.edu/seti/].

7 [http://www.bigear.org/].

8 [http://seti.uws.edu.au/].

9 I.S. Shklovskii and C. Sagan, *Intelligent Life in the Universe* (San Francisco: Holden-Day, 1966).

10 N. Kardashev, "Transmission of Information by Extraterrestrial Civilizations," reprinted in *The Quest for Extraterrestrial Life :A Book of Readings* ed. D. Goldsmith (Mill Valley, CA: University Science Books, 1980), 136–39.

11 "Declaration of Principles Concerning Activities Following the Detection of Extraterrestrial Intelligence" [available at http://www.seti-inst.edu/post-detection.html].

12 [http://astrobiology.arc.nasa.gov/workshops/1998/roadmap/index.html].

13 G.W. Wetherill, "Possible Consequences of Absence of Jupiters in Planetary Systems," *Astrophysics and Space Science* 212 (1994):23–32.

14 E. Jones, "Where Is Everybody?," *Physics Today*, August 1985, 11–13.
15 M. Hart, "An Explanation for the Absence of Extraterrestrials on Earth," *Quarterly Journal of the Royal Astronomical Society* 16 (1975): 128–35, reprinted in Zuckerman and Hart, *Extraterrestrials: Where Are They?* 1–8; D. Viewing, "Directly Interacting Extra-terrestrial Technological Communities," *Journal of the British Interplanetary Society* 28 (1975): 735–44.
16 B. Zuckerman, "Stellar Evolution: Motivation for Mass Interstellar Migrations," *Quarterly Journal of the Royal Astronomical Society* 26 (1985): 56–59, reprinted in Zuckerman and Hart, *Extraterrestrials: Where Are They?* 40–44.
17 R.N. Bracewell, "Communications from Superior Galactic Communities," *Nature* 186 (1960): 670–71; R.N. Bracewell, *The Galactic Club: Intelligent Life in Outer Space* (New York: Norton, 1976).
18 F. Tipler, "Extraterrestrial Intelligent Beings Do Not Exist," *Quarterly Journal of the Royal Astronomical Society* 21 (1981): 267–81.
19 NASA Office of Space Science Advanced Concepts Workshop, Robotic Interstellar Exploration in the Next Century, Caltech, July 28–31, 1998.
20 For general discussion of advanced propulsion techniques, see R.L. Forward, "21st Century Space Propulsion," *Journal of Practical Applications in Space* 2 (1991): 1–35; E.F. Mallove and G.L. Matloff, *The Starflight Handbook* (New York: Wiley, 1989).
21 S. Howe, "Antimatter (pbar) technologies" [available at http://www.roadrunner.com/~mrpbar/pbar.html].
22 C. Singer, "Settlements in Space and Interstellar Travel," in Zuckerman and Hart, *Extraterrestrials: Where Are They?* 70–85.
23 R.L. Forward, "Roundtrip Interstellar Travel Using Laser-Pushed Lightsails," *Journal of Spacecraft and Rockets* 21 (1989): 187–95.
24 C. Mileikowsky, "Cost Considerations Regarding Interstellar Transport of Scientific Probes with Coasting Speeds of About 0.3c" (paper IAA-94-655, 45th Congress of the International Astronautical Federation, Jerusalem, Israel October 9–14, 1994). For a lower estimate of the capital costs and further details of the technology, see G. Landis, "Small Laser-Pushed Lightsail Interstellar Probe," *Journal of the British Interplanetary Society* 50 (1997): 149–54.
25 The Lightcraft Project at Rensselaer Polytechnic Institute [available at http://www-aero.meche.rpi.edu/curriculum/tavd/].
26 B. Haisch, A. Rueda, and H.E. Puthoff, "Beyond $E=mc^2$", *The Sciences* (1994): 34, 26–31 (also available at http://www.jse.com/haisch/sciences.html); B. Haisch and A. Rueda, "An Electromagnetic Basis for Inertia and Gravitation: What Are the Implications for 21st Century Physics and Technology?," CP-420, Space Technology and Applications International Forum, ed. M.S. El-Genk (DOE Conference 960103, American Institute of Physics, 1998), 1443.
27 M. Papagiannis, "Are We Alone, or Could They Be in the Asteroid Belt?," *Quarterly Journal of the Royal Astronomical Society* 19 (1978):277.

CHAPTER 8

1 [http://www.ufomind.com/].
2 D.M. Jacobs, *The UFO Controversy in America* (Bloomington: Indiana

University Press, 1975); B. Steiger, ed., *Project Blue Book: The Top Secret UFO Findings Revealed* (New York: Ballantine, 1976).

3 University of Colorado, *Final Report of the Scientific Study of Unidentified Flying Objects*, E.U. Condon, director, and D.S. Gillmor, ed. (New York: Dutton, 1969).

4 J.E. Mack, *Abduction: Human Encounters with Aliens* (New York: Scribner's 1994).

5 P.A. Sturrock, director, "Physical Evidence Related to UFO Reports" (1998) [available at http://www.jse.com/ufo_reports/Sturrock/toc.html].

6 [http://www.csicop.org].

7 P. Davies, *Are We Alone? Philosophical Implications of the Discovery of Extraterrestrial Life* (New York: Basic Books, 1995), 131–38.

CHAPTER 9

1 G. Feinberg and R. Shapiro, *Life beyond Earth: The Intelligent Earthlings Guide to Life in the Universe* (New York: Morrow, 1980), 147.

2 A.G. Cairns-Smith, *Seven Clues to the Origin of Life: A Scientific Detective Story* (Cambridge: Cambridge University Press, 1985).

3 F. Drake, "Life on a neutron star," *Astronomy* 1 (1973): 4–8.

4 See the Tierra home page [available at http://www.hip.atr.co.jp/~ray/tierra/tierra.html].

5 T.S. Ray and J. Hart, "Evolution of Differentiated Multi-threaded Digital Organisms," in *Artificial Life VI Proceedings*, ed. C. Adami, et al. (Cambridge: MIT Press, 1998), 295–304 [also available at http://www.hip.atr.co.jp/~ray/pubs/pubs.html].

6 W.O. Stapledon, *Last and First Men: A Story of the Near and Far Future* (London: Methuen, 1930).

7 J.R. Searle, "Minds, Brains, and Programs," *Behavioral and Brain Sciences* 3 (1980): 417–24 (with appended critiques); J.R. Searle, "Is the Brain's Mind a Computer Program?" *Scientific American*, January 1990, 26–31; J.R. Searle, D.C. Dennett, and D.J. Chalmers, *The Mystery of Consciousness* (New York: *New York Review of Books*, 1997) (articles reprinted from the NYRB); for other critiques, see P. Churchland and P. Churchland, "Could a Machine Think?," *Scientific American*, January 1990, 32–37; S. Pinker, *How the Mind Works* (New York: Norton, 1997).

8 [http://www.research.ibm.com/deepblue/].

9 H. Moravec, *Mind Children: The Future of Robot and Human Intelligence* (Cambridge: Harvard University Press, 1988).

10 J. Lovelock, *Gaia: A New Look at Life on Earth* (New York: Oxford University Press, 1979).

11 R.R. Ruether, *Gaia and God: An Ecofeminist Theology of Earth Healing* (San Francisco: Harper, 1994).

12 E. Schrödinger, *What Is Life* (1994; reprint, New York: Cambridge University Press, 1967), 71.

13 L. Smolin, *The Life of the Cosmos* (New York: Oxford University Press, 1997), 156.

14 P.M. Churchland, *Matter and Consciousness: A Contemporary Introduction to the Philosophy of Mind* (Cambridge: MIT Press, 1988), 173.

15 E. Mayr, *The Growth of Biological Thought: Diversity, Evolution, and Inheritance*

(Cambridge: Harvard University Press, 1982), 53–67.
16 S. Harnad, "Levels of Functional Equivalence in Reverse Bioengineering," in *Artificial Life: An Overview*, ed. C. Langton (Cambridge: MIT Press, 1995), 293–302.
17 R.J. Cameron, "The Nature of Life" [available at http://ucsub.colorado.edu/~camerorj/cam.diss.chone.html].
18 J. Monod, *Le hasard et la nécessité* (Chance and necessity: An essay on the natural philosophy of modern biology) (New York: Knopf, 1971).

CHAPTER 10

1 In J. Horgan, *The End of Science: Facing the Limits of Knowledge in the Twilight of the Scientific Age* (New York: Broadway Books, 1996), 109.
2 R.H. Dicke, "Dirac's Cosmology and Mach's Principle," *Nature* 192 (1961): 440–41.
3 J. Gribbin and M. Rees, *Cosmic Coincidences: Dark Matter, Mankind and Anthropic Cosmology* (New York: Bantam Books, 1989).
4 J.A. Wheeler, "The 'Past' and the 'Delayed-Choice' Double-Slit Experiment," in *Mathematical Foundations of Quantum Theory*, ed. A.R. Marlow (New York: Academic Press, 1978), 9–48.
5 A.H. Guth and A.P. Lightman, *The Inflationary Universe: The Quest for a New Theory of Cosmic Origins* (New York: Perseus Press, 1997).
6 J. Glanz, "A Dark Matter Candidate Loses Its Luster," *Science* 281 (1998):332–33.
7 H. Martel, P.R. Shapiro, and S. Weinberg, "Likely Values of the Cosmological Constant," *Astrophysical Journal* 492 (1998): 29–40.
8 L. Smolin, *The Life of the Cosmos* (New York: Oxford University Press, 1997).
9 J. Gribbin, *In the Beginning: The Birth of the Living Universe* (New York: Viking, 1993).
10 B. Carter, *The Anthropic Principle and Its Implications for Biological Evolution*, Philosophical Transactions of the Royal Society of London, ser. A (1983), London 310, 347–63.
11 M. Livio, "How Rare Are Extraterrestrial Civilizations and When Did They Emerge?," *Astrophysical Journal* 511 (1999): 429–431.
12 J.R. Gott III, "Implications of the Copernican Principle for Our Future Prospects," *Nature* 363 (1993):315–19; J. Leslie, *The End of the World: The Ethics and Science of Human Extinction* (London: Routledge, 1996).
13 N. Bostrom, "Investigations into the Doomsday Argument," [available at http://www.anthropic-principle.com/preprints/inv/investigations.html].
14 [http://www.leaderu.com/real/ri9403/evidence.html].

CONCLUSIONS

1 [http://stardust.jpl.nasa.gov/top.html].
2 [http://www.coseti.org/].
3 T. Paine, *The Age of Reason* (1793; reprint, Secaucus, NJ: Citadel Press, 1974), 85.

Index